Slávka Kočanová
Ladislav Lukáč

Utilization of the recultivated area of the part landfill

AF138583

Slávka Kočanová
Ladislav Lukáč

Utilization of the recultivated area of the part landfill

Landfill gas and its use

LAP LAMBERT Academic Publishing

Impressum / Imprint

Bibliografische Information der Deutschen Nationalbibliothek: Die Deutsche Nationalbibliothek verzeichnet diese Publikation in der Deutschen Nationalbibliografie; detaillierte bibliografische Daten sind im Internet über http://dnb.d-nb.de abrufbar.

Alle in diesem Buch genannten Marken und Produktnamen unterliegen warenzeichen-, marken- oder patentrechtlichem Schutz bzw. sind Warenzeichen oder eingetragene Warenzeichen der jeweiligen Inhaber. Die Wiedergabe von Marken, Produktnamen, Gebrauchsnamen, Handelsnamen, Warenbezeichnungen u.s.w. in diesem Werk berechtigt auch ohne besondere Kennzeichnung nicht zu der Annahme, dass solche Namen im Sinne der Warenzeichen- und Markenschutzgesetzgebung als frei zu betrachten wären und daher von jedermann benutzt werden dürften.

Bibliographic information published by the Deutsche Nationalbibliothek: The Deutsche Nationalbibliothek lists this publication in the Deutsche Nationalbibliografie; detailed bibliographic data are available in the Internet at http://dnb.d-nb.de.

Any brand names and product names mentioned in this book are subject to trademark, brand or patent protection and are trademarks or registered trademarks of their respective holders. The use of brand names, product names, common names, trade names, product descriptions etc. even without a particular marking in this work is in no way to be construed to mean that such names may be regarded as unrestricted in respect of trademark and brand protection legislation and could thus be used by anyone.

Coverbild / Cover image: www.ingimage.com

Verlag / Publisher:
LAP LAMBERT Academic Publishing
ist ein Imprint der / is a trademark of
OmniScriptum GmbH & Co. KG
Heinrich-Böcking-Str. 6-8, 66121 Saarbrücken, Deutschland / Germany
Email: info@lap-publishing.com

Herstellung: siehe letzte Seite /
Printed at: see last page
ISBN: 978-3-659-81707-6

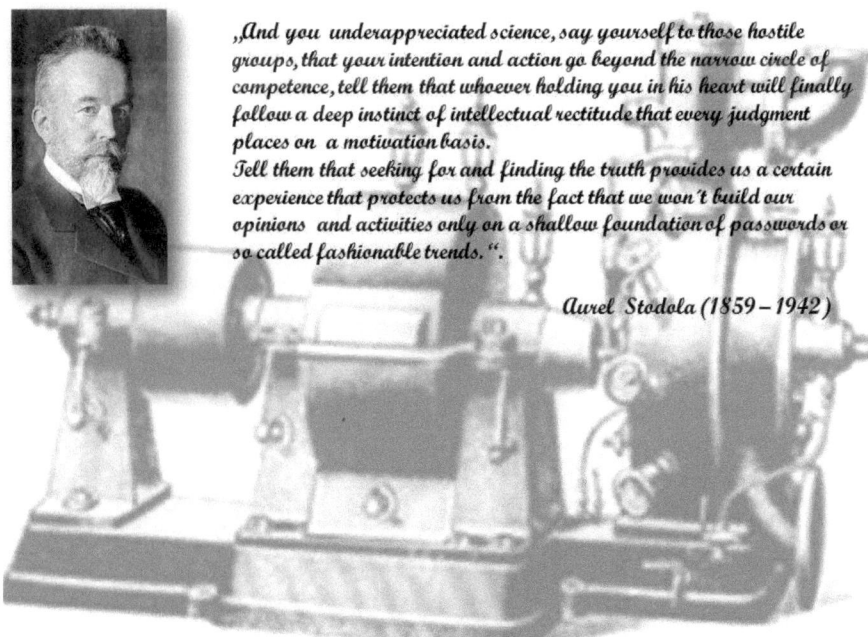

,,And you underappreciated science, say yourself to those hostile groups, that your intention and action go beyond the narrow circle of competence, tell them that whoever holding you in his heart will finally follow a deep instinct of intellectual rectitude that every judgment places on a motivation basis.

Tell them that seeking for and finding the truth provides us a certain experience that protects us from the fact that we won't build our opinions and activities only on a shallow foundation of passwords or so called fashionable trends. ''.

Aurel Stodola (1859 – 1942)

NOMENCLATURE

Dh_d	Daily sum of diffuse irradiation	$(kWh.m^{-2})$
T_{24}	Daily (diurnal) air temperature	$(°C)$
Di_d	Daily sum of diffuse irradiation	$(kWh.m^{-2})$
Es_d	Daily sum of specific electricity production	$(kWh.kWp^{-1})$
E_{share}	Percentual share of monthly electricity production	$(kWh.kWp^{-1})$
Es_m	Monthly sum of specific electricity production	$(kWh.kWp^{-1})$
Et_m	Monthly sum of total electricity production	$(kWh.kWp^{-1})$
Gh_d	Daily sum of global irradiation	$(kWh.m^{-2})$
Gh_m	Monthly sum of global irradiation	$(kWh.m^{-2})$
Gi_d	Daily sum of global irradiation	$(kWh.m^{-2})$
Gi_m	Monthly sum of global irradiation	$(kWh.m^{-2})$
PR	Performance ratio	$(\%)$
Ri_d	Daily sum of reflected irradiation	$(kWh.m^{-2})$
Sh_{loss}	Losses of global irradiation by terrain shading	$(\%)$

4

INTRODUCTION

The submitted monograph deals with the issue of landfill gas generation and its use in a combined heat and power system (CHP). Nowadays, landfilling, which is the least environmentally friendly, is the most common and cheapest way of waste disposal. The term of landfill site indicates a place in which waste is disposed. A legal landfill site, which is managed, is to ensure protection from ground water, soil and air pollution. Although a landfill site must be controlled and managed under the current legislation, it introduces plenty of potential hazards. The only benefit of landfills is that they can be seen as a possible source of landfill gas, which can be further effectively utilized as a fuel under certain conditions.

The monograph addresses the total quantities of waste generated, its disposal and recovery in the Slovak Republic at the end of 2013. Moreover, it also provides brief characteristics of landfill gas generation in municipal waste landfill sites, the results of calculations of the theoretical amount of landfill gas generated on the existing landfill sites in the Slovak Republic and theoretically possible electricity and heat generation by a CHP system.

The fifth chapter focuses on simulating the landfill gas generation. A model, input variables and the simulation results are described in this chapter.

The experimental part of the book consists of experimental measurements carried out at the Úholičky landfill. The main objective was to find out the percentage composition of landfill gas in the individual wells and the critical number of wells.

The sixth chapter describes a proposal of utilization of recultivated area of the Úholičky landfill, economic evaluation of the proposal, which includes budget of particular items of the Photovoltaic Power Plant (FVE) of the landfill site, and estimated revenues and payback period of the investment for the proposed output. In addition, the chapter deals with the expected annual electricity generation at the landfill, that is estimation of profitability of the photovoltaic (FV) system in the initial phase of operation.

The latest report on how Member States manage their municipal waste (MW) shows that there are significant differences across the EU. The EU gave Slovakia the lowest mark mainly due to a low fee for MW disposal at landfills, non-existent waste prevention programme and it recommended to increase a rate of recycled MW.

The above mentioned data and facts indicated that landfilling is the most common method of waste treatment in Slovakia and it will likely remain so for some time in the future. I find it important to pay attention to the topic of landfills and energy recovery of landfill sites.

1. THE WASTE STATISTICS IN THE SLOVAK REPUBLIC

According to the reports of the Statistical Office of the Slovak Republic, Slovakia generated 8114 592.5 tons of waste in 2013. Of the total amount of waste generated, a weight of 3 736 241 tons was landfilled, 59 048.5 tons was incinerated without energy recovery and a weight of 455 782.6 tons of waste was disposed by means of another method. Of the total amount of waste generated in the given year, a weight of 135 341.3 tons of waste was energy recovered and 2 057 327.1 tons of waste was material recycled. The exact quantities of disposed and recovered waste are in Table 1.

Table 1 Quantities of disposed and recovered waste in 2013 [1]

	landfilling	3 736 241,0
Disposed of waste in tonnes	combustion without energy recovery	59 048,5
	another disposal method	455 782,6
	materially	2 057 327,1
Recovered of waste in tonnes	energy	135 341,3
	composting	619 847,6
	other recovery method	898 784,7

In Slovakia, the total waste recovered accounted for 27.78 %, of which 13.59 % of waste was material recycled, 35.83 % was energy recovered, 26.96 % was composted and 23.62 % was treated by other methods. The waste disposed amounted to 70.56 %, of which 97.64 % of waste was landfilled.

Compared with 2012, an annual increase of waste was by approximately 13.7 % in 2013. A significant share in growth of waste is represented by other sorts of waste (7,750,050.87 tonnes in 2013 compared with 6,548,981.86 tonnes in 2012). In 2013, Slovakia generated 1,744,428.65 tonnes of municipal waste (MW) altogether, which represented about 322 kg of municipal waste per capita. It represented a decrease by 1 kg compared with 2012. The trend of municipal waste generation is more or less steady in the long run. The Bratislava region achieved the largest share in municipal waste generation per capita again. It is proportional to the economic strength of the region. Industry, which contributed 30% of the total waste generated, is the largest waste producer and is followed by construction industry with about 25% of waste generated; the remaining approximately 11% was generated by electricity, gas and steam suppliers [1,3].

The Prešov region generated the least waste per capita and the Bratislava region the most. The Prešov region generated a weight of 242.16 kg of municipal waste per capita, which was by 80.08 kg of waste less than the amount of waste generated per capita in Slovakia. The Bratislava region generated about 104.12 kg of waste more than the amount generated per capita in Slovakia [1].

Landfilling has been one of the simplest and most common methods of waste disposal for several years. Any waste which cannot be material recycled, recovered and/or separated is landfilled. The dominant part of municipal waste treatment also consisted of landfilling in 2013. However, a share in waste landfilled decreased below 70% for the first time that year [3].

A landfill is a facility or a place in which waste is deposited on the surface or beneath the earth's surface. It is desirable to comply with safety, health and environmental aspects. Landfill is the last link in the chain of waste dumping [2,4].

1.1 LANDFILLS IN THE SLOVAK REPUBLIC

A place intended for waste disposal is considered to be a landfill. This is a place where the waste is deposited on or beneath the earth's surface. The purpose of the landfill is to dispose waste in such a way that there would be an absence of oxygen and the waste would be isolated from ground water and kept in a dry place.

Landfills are divided into the following categories:

- ➢ Non-hazardous waste landfills
- ➢ Hazardous waste landfills
- ➢ Inert waste landfills [2].

The basic requirement for building a landfill, which is laid down by law, is that the landfill must be sealed (by an artificial seal or geological barrier) in such a way that the protection of ground, surface water and the total soil protection would be ensured [2,5]. In Fig. 1, there is a placement of landfills on which non-hazardous, inert and hazardous waste is disposed in Slovakia at the end of 2013.

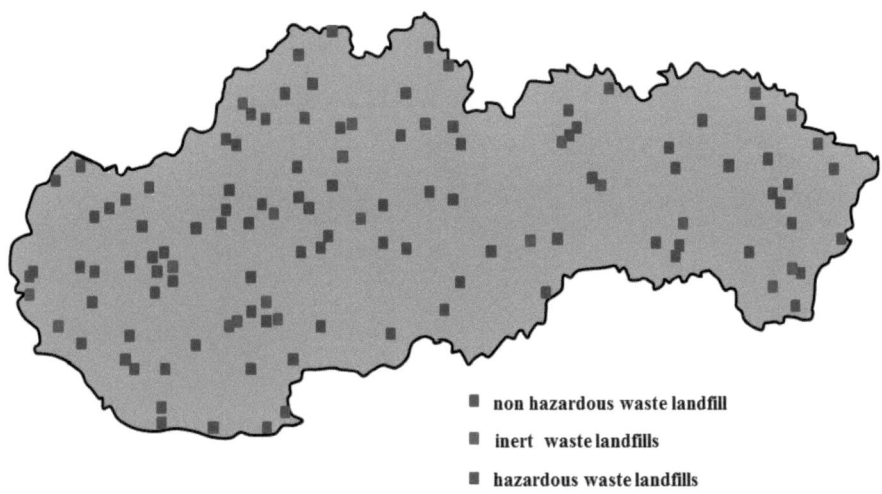

Figure 1 Placement of non-hazardous, inert and hazardous waste landfills in slovakia[1]

There were 124 controlled landfills in the Slovak Republic at the end of 2013. Landfills, which were classified by regions, in Slovakia on December, 31st, 2013 are shown in Fig. 2. The non-hazardous landfills accounted for 96; the other 18 landfills were used to deposite inert waste and the remaining 10 were used for disposal of hazardous waste [1].

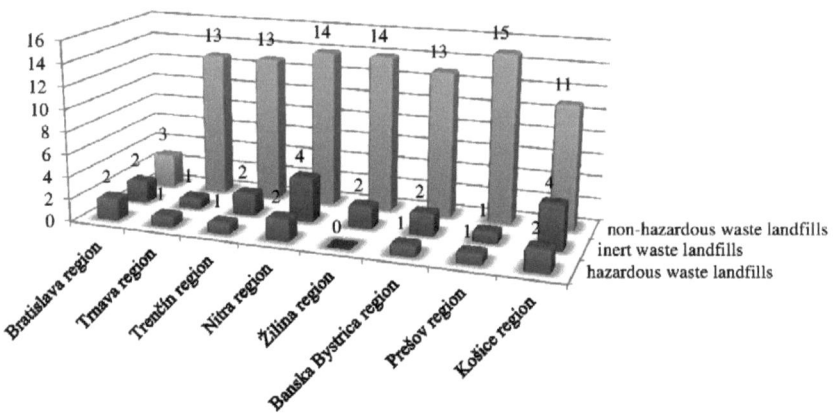

Figure 2 The number of non-hazardous, inert and hazardous waste landfills in Slovakia in 2013

Compared with the previous years (e.g. the total number of landfills accounted for 151 in 2007), the landfills have showed a downward trend in our country if we consider the legal (controlled) landfills. [1]

Placing waste at landfills allows waste recovery in relation to obtaining energy in the form of biogas (landfill gas) to a limited extent. Each landfill should be equipped by a collection system for capturing landfill gas (LFG) and leachate. Landfill gas standardly is captured from the beginning of the operation of the landfill to the period of 10 years after it has been decommissioned. The composition of LFG depends on the composition of the waste received in the landfill and the length of its disposal [10].

2. LANDFILL GAS GENERATION IN MUNICIPAL WASTE LANDFILLS

Gas generation in the municipal solid waste landfills has been monitored since the early 70s of the 20th century. In general, the monograph by Farquah and Rovers, who were the first to describe a sequence of biomethanation phases as the landfill's environment changes with time, is considered a milestone of the expansion of information on and knowledge of the gas generation in landfills [8].

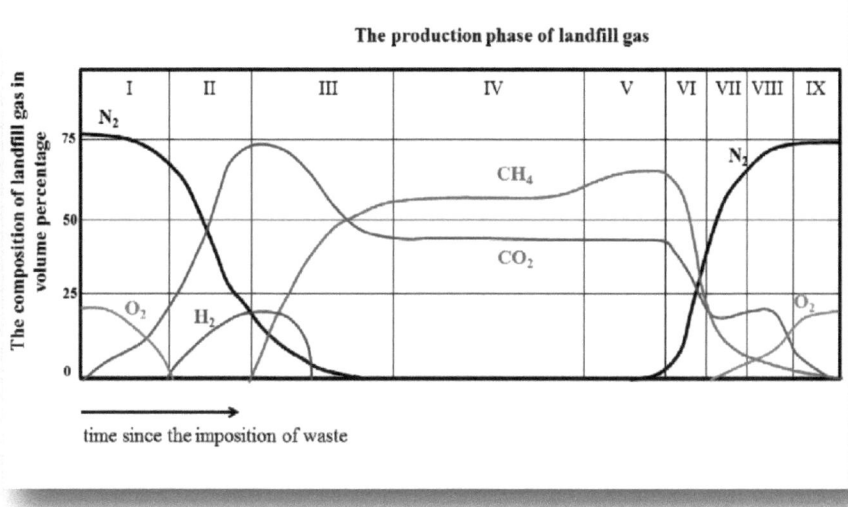

Figure 3 The course of landfill gas composition at continuous biomethanation [8-15]

Biomethanation in landfills cannot be described by generally valid time dependence, which is why there is not a fixed timeline attached to the concentration curves of various gas components shown in Fig. 4 [8-15].

The progress of methanogenic processes is not entirely obvious even if they are completely spontaneous. For successful biogas generation, it is necessary to meet the following conditions:

➢ *Oxygen is not supposed to have access to the landfill*

The landfill must therefore be sufficiently deep, compacted and air is not supposed to have any access to the landfill through any drainage.

➤ *The waste must be sufficiently moist*

Moisture requirement is as important as the requirement for absence of oxygen. The anaerobic decomposition processes cannot take place in the surroundings with insufficient moisture; even the methanation processes, which have already started, stop when there is not sufficient moisture content.

➤ *The waste mustn't contain any bactericidal or other toxic or inhibitive substances.*

For example, waste discharged from wood manufacturing treated with anti-fungal and anti-decomposition agents is very difficult to degrade

There are many other conditions for the quick start of methanation processes. However, they are of lower importance and their failure will not cause a complete halt of biological processes [6].

2.1 THE MAJOR AND MINOR LANDFILL GAS COMPONENTS

The major landfill gas components include methane and carbon dioxide. The content of majority of the other gases is lower by more than one grade.

The major components of landfill biogas significantly differ from those of the reactor biogas. The landfill is not an ideal gastight vessel in contrast to the reactor; the diffusion processes and the effects of changing barometric pressure will result in the gas containing the diluted residues of the reacted air or comprising of a fraction of unchanged air. Besides methane and carbon dioxide, landfill gas contains oxygen when compared with reactor gas; it can also consist of argon and unreacted oxygen [7,8].

The initial decomposition of biodegradable matter of the waste begins during the waste collection and transport phases in the form of the hydrolytic aerobic processes. After the waste has been transported to the landfill and the layer of waste has been made compact, relatively rapid depletion of oxygen occurs in the recently compacted layer and the aerobic processes begin to shift to the anaerobic ones. Given that there is not a perfect contact of the phases ensured in the landfill, the processes of acidogenesis and methanogenesis proceed more slowly and, at first, without mutual progress. Therefore, a state of well-developed acidogenesis temporarily functioning without methanogenesis can be observed in the newest parts of the landfills. The beginning of the methanogenic processes is relatively slow and it mainly requires a change of pH and the complete removal of oxygen. In the initial period (acidogenic phase), the unusual composition of the gases extracted from the relevant zones can be observed. The gas may contain over 50% of carbon dioxide and trace quantities of hydrogen can be found in it [9]. Trends in landfill gas composition during the particular phases are shown in Fig. 3, page 11.

The biomethanization process inside the landfill along with acidogenesis is slowly getting balanced and these interims of different lengths are indicated as an unstabilized methanogenic phase. The term "unstabilized" means that the biodegradation processes of the landfill are influenced by many more factors. Progress of methanization processes may be limited or virtually halted if there is a limited water supply. Inappropriate placement of drainage system can supply the landfill with air, which may also lead to limitation or even a complete halt of methanogens development and the restoration of the acidogenic phase.

Therefore, we can conclude that landfill gas is mainly a three-component mixture containing of CH_4, CO_2, N_2 in terms of the major components. The typical examples of the landfill gas composition in various development states of the landfill body are in Table 2.

Table 2 Gases composition in landfills, in volume per cent [7,9]

Phase	CH$_4$	CO$_2$	O$_2$	N$_2$	H$_2$
Acidogenic	0	80	0	18	2
Methanogenic unstabilized	20	64	0	16	0
Methanogenic unstabilized	40	55	0	5	0
Methanogenic stabilized	62	37	0	1	0
Methanogenic unstabilized (landfill overloaded exhausting)	47	33	0	20	0
Aerobic (long-term overloading of the landfill)	40	27	3	30	0

The first changes of landfill's state in relation to aerobic process are always apparent in the form of a decrease in methane and an increase in nitrogen oxides. The long-term overloading of the landfill's body leads to increasing quantities of oxygen content in the gas together with a very rapid decrease in methane and a subsequent increase in the nitrogen content.

The course of the particular landfill gas component concentration at intensive pumping out obtained in practice can be seen in Fig. 4.

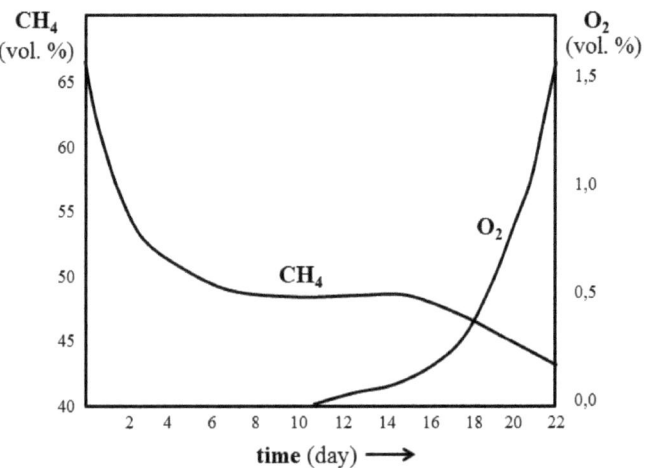

Figure 4 Changes in landfill gas composition at long-term extraction [7]

14

The initial velocity of a decline in the methane concentration from the stationary state leads to the apparent stabilization of components over time. Tracking the changes in nitrogen and oxygen indicates a future rapid decrease that occurs after the aerobic process takes place in the area around the well, which is damaged by extraxtion. Restoring the landfill to the acidogenic phase does not only affect the gas composition [6].

2.1.2 THE MINOR LANDFILL GAS COMPONENTS

Chemical composition of the minor landfill gas components is very wide compared with the major components. The chemical substances, which were identified in the landfill gas or biogas and were determined in hundreds of milligrams per cubic meter, are calculated in hundreds for plenty of groups and types of derivatives:

- Aliphatic hydrocarbons
- Alicyclic hydrocarbons
- Aromatic hydrocarbons
- Alcohols and Thiols
- Alcohols and Ketones
- Carboxylic acids
- Esters
- Halogenoalkanes
- Amines
- Furan and its derivatives.

In tables 3 and 4, there are some selected examples of concentration range of the minor hydrocarbon components in the landfill gas. The challenge of analysis of such a wide spectrum of organic compounds lies in accurate and correct identification and proper resolution of the chromatographic peaks in most of the cases [9].

Table 3 The selected examples of concentrations range of the minor hydrocarbon components in the landfill gas according to different authors, in mg.m^{-3} [17,18]

Minor impurities	Rettenberger	Young
propane	1,4-13	< 2-100
butane	0,3-23	10-560
pentane	0-12	0,8-613
hexane	3-21	0,5-76
heptane	3-37	2,6-47
octane	0,15-80	5,7-99
nonanes	0,05-400	10,2-270
decanes	0,2-137	3,8-257
undecanes	7-48	< 2-86,4
dodecanese	2-4	< 0,5-6,4
tridecan	0,2-1	not available
toluene	0,2-615	18-197
xylenes	0-383	7,9-139
ethylbenzene	0,03-7	3,6-49

Table 4 Some typical minor components of landfill gas in the form of derivates, in mg.m^{-3} [17,18]

Derivates	Concentration
methanol	2-210
ethanol	16-1450
1-propanol	4,1-630
2-propanol	1,2-73
1-butanol	18-626
acetic acid	< 0,06-3,4
butyric acid	< 0,02-6,8
butyl acetate	60

The analyses of minor content of hydrocarbons are nearly worthless for practical purposes since these substances contribute to the total energy content of landfill gas to a very little extent. Higher calorific value of gas is virtually given by the methane content. If all the combustible trace compounds were added and their impact on the overall higher calorific value of the gas was calculated, we find out that the increase in the value of higher calorific value is lower than 1-2%, even if the amount of contents is on the upper data limits.

The minor components occurring in the landfill gas can be divided into natural substances, derived from decomposition of natural materials and components

16

originating in artificial materials. The origin of many chemical compounds, which are found in landfill gas, is often very difficult or impossible to determine.

It is important to pay attention to the technically significant compounds of gas in another overview of the influences, effects and behaviour of the minor components of landfill gas [6].

2.1.3 THE MINOR LANDFILL GAS COMPONENTS

When assessing the types of landfill biogas, it is necessary to compare the chemical analyses of the collected gases. An approximate converting of the results of chemical analysis to the formal components is suitable for fast and transparent comparison. The four basic components of the gas (CH_4, CO_2, O_2 and N_2) are converted to four components which correspond to the formally determined chemical composition.

The gas extracted from the landfill body does not necessarily come from the activity of methanogenic microorganisms. In most of the cases, a sample of the gas taken from the landfill is a mixture of gases of different origin.

The main landfill gas components are essentially as follows:

1) **"Clean biogas"** - a model gas ($CH_4 + CO_2$) of standard composition corresponding to the higher content of methane , which was obtained by anaerobic digestion;

2) **"Unchanged air"** - infiltrated unchanged air;

3) **"Stuffy air"** - a gas which was generated as a residual compound after the oxygen from the air had been consumed; it consists of N_2 and CO_2;

4) **"Gas from acidogenic processes"**- a gas that has a surplus of CO_2 [6].

2.2 MUNICIPAL WASTE LANDFILL DEVICES USED FOR DEGASSING

Degassing of decommissioned and operated landfills is necessary for safety reasons of the further usage of the landfill body, the surroundings and excluding negative impact on the environment [6]. When a landfill is not covered with any surface sealing material, the natural degassing takes place by gas circulation and oxidation of methane. To avoid uncontrolled movement of gas inside the landfill and its release to the environment, gas capturing facilities are built and barriers preventing gas from leaking to the surroundings of the landfill also called degassing systems.

The term "degassing system" essentially means construction-technical and technological equipment located inside the landfill and within its body. The equipment also includes pumping systems and landfill gas usage or disposal devices [7, 21].

The degassing systems classified by the method of gas drainage are as follows:

- ➢ Active
- ➢ Passive.

2.2.1 ACTIVE LANDFILL GAS CAPTURING

The active degassing systems (Fig. 5) extract gas from the landfill bodies by means of various types of gas blowers. They are designed either with lower or upper gas outlet. The basis of the landfill gas collection network consists of boreholes or continuously excavated wells. These vertical parts of the system are built in the form of 0.8 to 1.0 meter wide coarse gravel fillings with a central perforated collecting casing pipe, which usually interconnects all the parts of a vertical borehole [7, 11].

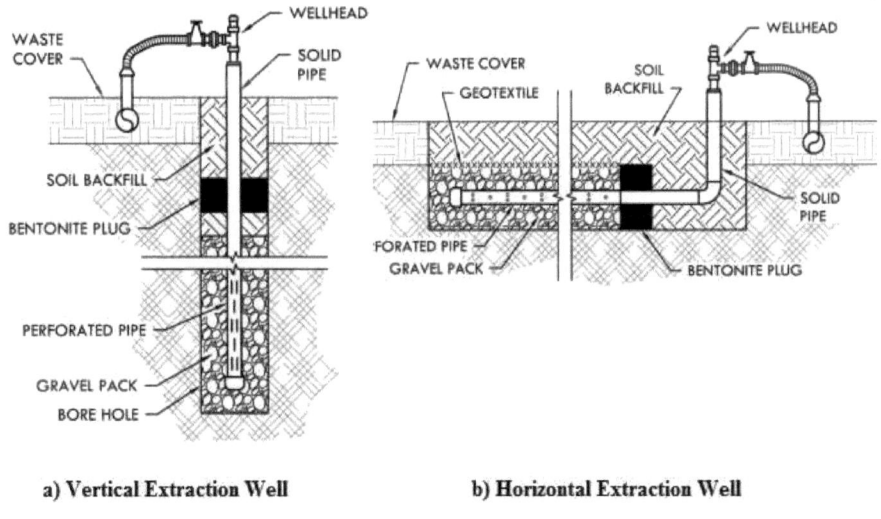

a) Vertical Extraction Well   **b) Horizontal Extraction Well**

Figure 5 A model cross-section of a gas well [10]

2.2.2 PASSIVE LANDFILL GAS CAPTURING

This method is based on utilisation of the pressure of the gas to ensure that the flow is directed correctly. Another option is that under pressure is created in order to extract the landfill gas, which is subsequently utilised or disposed of. Permeable layers by means of holes or apertures are created in the landfill site in order to make the gas capturing process possible. Finally, the difference between the landfill and the environment pressures leads to a gas leak through these openings. The passive gas capturing system is sized based on the quantities of the gas contained in a landfill, however, its amount and quality may vary to some extent in practice [7, 21].

3. THE THEORETICAL CALCULATION OF METHANE GENERATED FROM MUNICIPAL WASTE

When making a decision about how to use the landfill gas at the municipal waste landfills, the knowledge of gas generation is an important parameter for understanding the process. The specific gas production, thus potential of the landfill gas, can be defined as the total volume of gas produced from a particular quantity of waste in the conditions of the landfill. There are plenty of relationships for determination of the production amount of landfill gas, which are based on two principles, biochemical and chemical-physical models. Theoretical calculations can be used to determine the amount of methane generated from municipal solid waste. This issue has been addressed since 1930.

In the Slovak Republic, all registered landfills are registered and managed by the State Geological Institute of Dionýz Štúr. The landfills register in Slovakia has been kept since 1992 and its data (landfill operator, name, size, the year of establishment of the landfill, the year of decommissioning and many others) is regularly updated.

The theoretical quantities of landfill gas obtained from the existing landfills in Slovakia, theoretically possible electricity and heat generation by CHP were calculated based on the theoretical calculations.

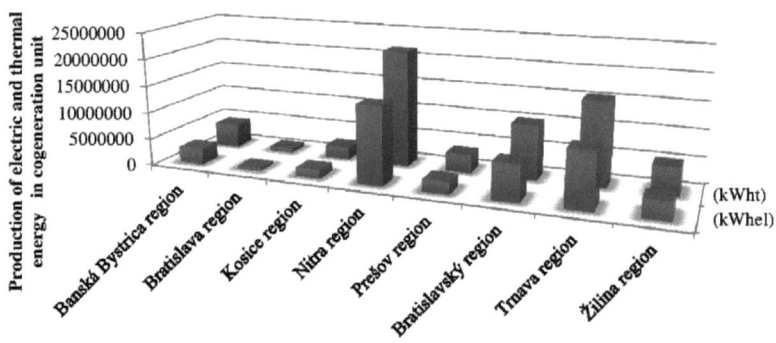

Figure 1 The theoretically possible electricity and heat generation in Slovakia; classified by regions

20

3.1 ENERGY UTILISATION OF LANDFILL GAS

Landfill gas must be extracted from all the landfills, in which biodegradable waste is deposited. The condition is that the gas must be generated in the technically processable quantities. The collected LFG must be treated and utilised for heat and power generation. The collection, treatment and, last but not least, utilisation of LFG must be done without negative impacts on human health or the environment [2]. The LFG collected from non-hazardous waste landfills is basically used for combined heat and power generation (CHP) [25, 26].

3.1.1 COMBINED HEAT AND POWER AND ITS ADVANTAGES AND DISADVANTAGES

Cogeneration, i.e. combined heat and power (CHP), is a highly efficient energy system, which generates electricity and heat using a single fuel source [27]. It mainly provides the environmental and economic profit as otherwise waste heat turns into a useful energy source [5]. An overview of the advantages and disadvantages of combined heat and power is listed in Table 5 [27].

Table 5 An overview of advantages and disadvantages of combined heat and power generation

Advantages	Disadvantages
➢ There is a 40% fuel savings compared with traditional technologies	➢ Noise
➢ fuel utilization to 80-85% (65-70% on heat energy and 30-35% on electrical energy)	➢ High investment costs
➢ Heat and electricity are produced from renewable energy sources	

3.1.2 ELECTRICITY GENERATED FROM LANDFILL GAS

Electricity generation is still the most common method of energy utilisation of LFG [30]. Electricity can be generated by combustion of landfill gas in an engine, gas turbine or micro-turbine [26, 32]. The advantages and disadvantages of these options in electricity generation using LFG are shown in Table 6.

Table 6 An overview of advantages and disadvantages of the landfill gas energy utilization technological facilities for electricity generation [28]

Advantages	Disadvantages
Internal combustion engine	
➢ Efficiency increases when waste heat is recovered ➢ High efficiency compared with gas turbines and microturbines ➢ Relatively low cost on a per kW installed capacity basis when compared with gas turbines and microturbines ➢ Can add or remove engines to follow gas recovery trends	➢ Economics may be marginal areas with low electricity costs ➢ Relatively high air emissions ➢ Relatively high maintenance costs
Gas turbine	➢ Economics may be marginal areas with low electricity costs
➢ Efficiency increases when heat is recovered ➢ Low nitrogen oxides emissions ➢ Cost per kW of generating capacity drops as the size of the gas turbine increases, and the efficiency improves as well	➢ Efficiencies drop when the unit is running at partial load
Microturbine	
➢ Low nitrogen oxides emissions ➢ Ability to add and remove units ➢ Can function with lower percent methane ➢ Requires lower gas flow	➢ Economics may be marginal areas with low electricity costs

3.1.3 ELECTRICITY GENERATED FROM LANDFILL GAS

Direct combustion is the easiest and often the most cost-effective way of LFG utilisation for heat generation [32]. The most common method of the direct use of LFG is combustion of the gas in boilers, in which the LFG is used as a fuel serving for steam generation; the steam is used for heating. The direct LFG utilisation is used for heating greenhouses if they are close to the landfill. Landfill gas is an ideal fuel

22

for infrared heating [26]. The advantages and disadvantages of LFG utilisation for heat generation are pointed out in Table 7[28],[32] .

Table 7 An overview of advantages and disadvantages of the landfill gas energy utilization technological facilities for heat generation [28]

Advantages	Disadvantages
Boiler, dryer and kiln	
➢ Does not require large amount of LFG and can be blended with other fuels	➢ Cost is tied to length of pipeline; energy user must be nearby
➢ Uses maximum amount of recovered gas flow	
➢ Limited condensate removal and filtration treatment is required	
Infrared heater	➢ Seasonal use may limit LFG utilization
➢ Does not require a large amount of gas	
➢ Can be coupled with another energy project	
➢ Relatively inexpensive	➢ High capital costs
Leachate evaporation	
➢ Good option for landfill where leachate disposal is expensive	

The total useable quantity of landfill gas in the boiler depends on its size. Large boilers are intended for supplying industrial premises and the technological processes which take place there by means of steam. Small boilers are designed for heat generation in edifices and buildings such as hospitals, schools [32].

3.2 ENERGY USE OF LANDFILL GAS IN OPERATION OF LANDFILL SITES IN SLOVAKIA

Landfill gas can very well be used in thermal plants of small outputs for the decentralized supply of the defined sites with heat and electricity. However, the gas is the most often combusted in a combined power and heat generation system (CHP) in Slovakia.

In Slovakia, eight facilities take advantage of the technology of landfill gas combustion in a CHP unit (Tab. 8).

Table 8 Energy utilisation of landfill gas in a CHP system in Slovakia

Locality	Operator	In operation since	Energy utilization	Installed power capacity
Landfill MSW A.S.A Zohor	TENERGO Brno, Inc.	Sep. 2008	Electricity generation 2400 MWh/year	0,32 MWe
Luštek - Dubnica nad Váhom	MAEN SK, Ltd.	2010	Electricity generation - yearly	0,150 MWe
Nový Tekov	MAEN SK, Ltd.	2009	Electricity generation 1 720 MWh per year	0,270 MWe
Landfill MSW Žilina, Považský Chlmec	Terrasystems, Ltd.	Feb. 2009	Electricity generation 2500 MWh/year	0,349 MWe
Landfill MSW SEKOLÓG Brezno	MAEN SK Ltd.	Jan. 2012	Electricity generation 1090 MWh/year	0,15 MWe
Landfill MSW Zvolenská Slatina	MAEN SK, Ltd.	2008	1079 MWh/year	0,15 MWe
Landfill MSW Banská Bystrica	MAEN SK, Ltd.	2008	Electricity generation 1262 MWh/year	0,15 MWe
Landfill MSW Žakovce - Úsvit	MAEN SK, Ltd.	2009	Electricity generation 1158 MWh/year	0,15 MWe

The system is based on landfill gas extraction from the body of the landfill at the under pressure conditions. The degassing system of the landfill consists of the extraction boreholes, pipeline networks, a condenser manhole and a pumping station. The landfill gas system is managed by operators whose container devices for gas collection and combustion in a CHP system are placed in the landfill.

4. CHARACTERISTICS AND A BRIEF DESCRIPTION OF A SELECTED MUNICIPAL WASTE LANDFILL

The large-scale landfill in Úholičky, which is located to the north of Prague, was founded by Regios, a.s. as a waste repository of the categories "SMW" (solid municipal waste) and "O" (the other types of waste that do not have any dangerous properties) together with the sub-category S-OO3 by the leachability class. In the present day, it serves for a catchment area of 25 km and the waste is transported by a total of 62 municipalities to the site.

The company provides comprehensive services in the waste management (municipal waste transport, operation of civic amenity site, separate collection, wood waste recycling, rehabilitation work, project preparation, a production line for solid alternative fuels and cleaning of the road communications by machines). Regios, a.s. is a subsidiary of A.S.A., spol. s.r.o. It is estimated that there are about 80 to 100 passages of trucks per day. A view of different zones of the landfill is in Fig. 6.

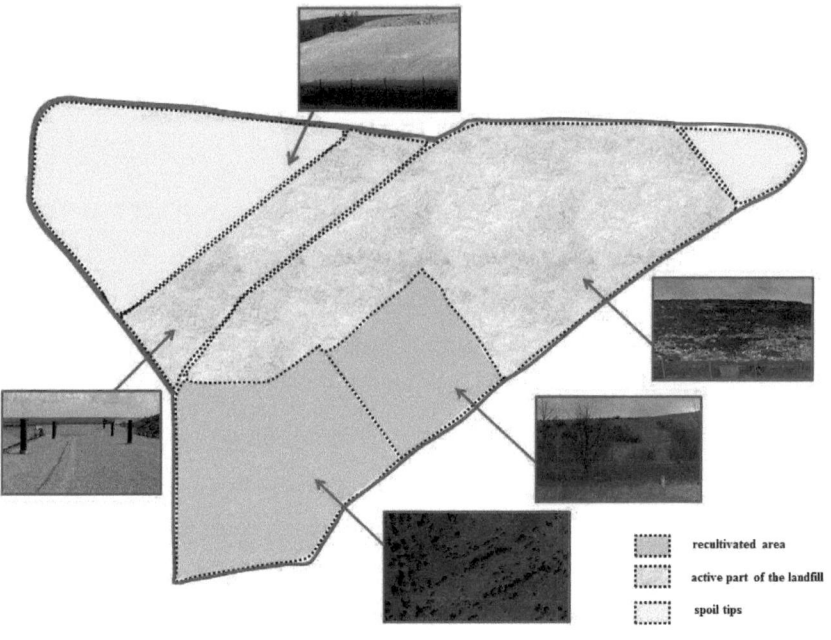

Figure 6 A view of different zones of the landfill site

The landfill site in the administrative area of Úholičky is a facility used for removal of other types of waste. The construction of the first phase of the landfill body was launched in 1994. Currently, phase 4 is already under construction. The trend in the total volume of the landfill in December, 2013 is visible in Table 9. Various phases of the landfill site including the recultivated areas are marked and shown in Figure 6 at the end of 2013.

Table 9 The total volume of the landfill

The landfill body	Landfill volume in m³
The total landfill volume to the design of the dome	3 107 000
The total volume of landfill I., II. Sectors 6-13 stages III. and IV. stages to a height	3 806 800
The volume of landfill only IV. stage in altitude difference of dome	1 379 061
The volume of landfill only IV. stage in the extent realized of the stages construction	513 747
The actual volume of landfill operational part I., II. A 6-13 sectorsIII. Sector III. stages and sectors 1-2 IV. stages (transported waste without recultivated)	2 752 201
The total volume of soil for recultivation	77 755
Total area already transferred by the reclamation	55 756

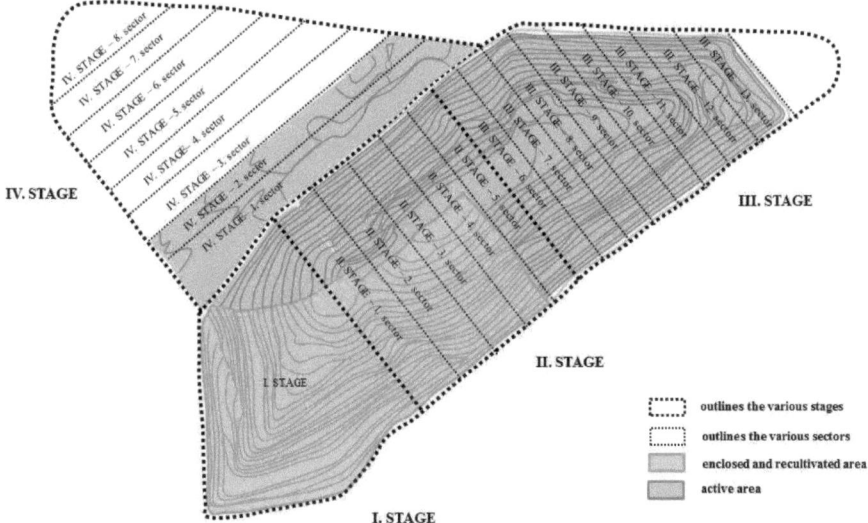

Figure 7 The landfill site of MW in Úholičky with various phases of the site including the recultivated areas marked at the end of 2013

The inclination of the slope of the landfill body is 1: 2.5 and the inclination on the top of the site is 1:10. The maximum waste elevation is 330 meters above sea level, which is in accordance with the terms of the planning permission. There is a service road of 3.5 m in width in the slope, which enabled the vehicles to drive up to the top of the site body. The road joins the existing road in the recultivated area.

4.1 WATER DRAINAGE SYSTEM IN THE AREA OF THE LANDFILL IN ÚHOLIČKY

The bottom of the landfill is shaped in such a way that it enables water to be drawn off easily into the drainage pipe system. The entire bottom is covered with a layer of drainage gravel of 300 to 500 mm. The dump water is drained from each section through a drainage system to a drain water shaft. Leachate (waste water from decomposition processes and rainwater flowing through the landfill body) is drained into a two-chamber sedimentation shaft, which includes a leachate service station, from there and further to a leachate tank with no outlet (3000 m^3). The tank is designed as impermeable (mineral insulation of 2 by 200 mm, covered with a film of 300 mm in width, fractions of 32 to 63 mm). There is a control drainage system connected to the control shaft of the leachate tank underneath the leachate tank. In case of any surplus, the accumulated leachate can be transported to a contractual sewage treatment plant (STP).

The dump water is used to irrigate the inner content of the recultivated area; the amount of reclaimed water is about 18,000 m^3 per year.

The rainwater management of the landfill uses the drainage channels and a rainwater sewage system. The rainwater is drawn off the recultivated area of the landfill through the concrete drainage channels and then it is pumped into the tank of clean water. Furthermore, the rainwater is released into the local watercourses.

A surface water cesspit, which will be used to collect rainwater for further utilization for irrigation of the landfill surface, will be built at the landfill in Úholičky in the future.

The recirculation of the dump water is crucial for proper development of biodegradation processes, limitation of dust and for the future stability of the landfill body as well. Since the area has lower total annual precipitation, the dump water must be continuously recycled and supplied to the surface of the landfill. Leachate from the landfill site is captured and collected in a central tank situated in the southern part of the site.

4.2 THE LANDFILL GAS TREATMENT AT THE LANDFILL SITE IN ÚHOLIČKY

The landfill is equipped with a gas drainage system. In total, there are currently 79 collection gas wells at the site, of which 28 are located in the recultivated area and are equipped with end elements connected to the biogas exhaust system (co-called top exhaust). There are gas pipelines of PEHD 90 to 160 mm in diameter laid on the recultivated layer.   Stand-alone connections, which are laid on a surface of the proper inclination and connected to the gas pipelines in the site body, are routed from the collection wells along the surface of the recultivated layer. The gas pipelines are connected to a collector of biogas PEHD D225 kept at the heel of the landfill site.

The gas is discharged through the main gas pipeline to the compressor biogas fuel station BSG 160 made of steel. The walls, floor and ceiling are insulated with a layer of basalt ORSIL. Two fans provide ventilation in case of emergency. The technological part of the device makes up of a unit of output of 2 by 300 $m^3.hod^{-1}$ and a stand-alone section for measurement and control.

The stand-alone part provides a continuous measurement and regulation of the extracted amount of biogas (the contents of CH_4 and O_2, temperature, pressure). The average amount of extracted landfill gas from the landfill in Úholičky was about 1,021,057.29 m^3 in seven years. The exact values of the amount of the extracted landfill gas and a share of methane in the landfill gas are illustrated in Table 10.

Table 10 The amount of draw off landfill gas together with a share of methane contents

A year	The amount of drawn off landfill gas (m³)	Proportion of methane in landfill gas (% CH₄)
2007	1 088 208,0	49,30
2008	1 437 413,0	46,47
2009	951 496,0	44,20
2010	95 675,8	41,70
2011	1 077 922,0	45,38
2012	1 033 834,0	56,70
2013	1 462 859,0	55,40

The pipeline continues further from the service station and it is routed to the flare stack of the residual landfill gas (LG) VTP BIO 600 (Fig. 8) and the cogeneration unit MOTORGAS TBG 520 (Fig. 9).

The structure of the flare stack used for burning residual LG is self-supporting with the outer stainless steel casing. The combustion chamber is made of refractory metal and insulated with ceramic fiber resistant up to temperature of 1250 ° C. It is also equipped with two outlets for flue gas collection and places for the location of the temperature sensor which are distributed evenly along the height of the casing.

The inner part of the flare stack is protected against atmospheric conditions by a circle cover. The combustion air is regulated by two louver dampers that are controlled by an actuator based on the temperature course in the combustion chamber. The injector burner runs on biogas. Flame supervision is done by a UV flame probe.

The main gas supply is equipped with flame arrester fuse fittings, a temperature sensor and a quick solenoid valve. The control system is placed in the switchboard at the burner; the superior control circuit is placed in the container for measurement and control. The flare stack VTP can be used to combust landfill gas as well as biogas from sewage treatment plants and sugar industry waste water in an environmentally friendly way.

Figure 8 Flare stack for combustion of residual landfill gas VTP BIO 600

The cogeneration unit MOTORGAS TBG 520 is mounted in a stand-alone transportable container made of steel, which is fitted with anti-noise elements. The cogeneration unit is of modular type of construction.

The installed cogeneration unit contains a gas engine of Waukesha L36 GLD type produced by an American company, Waukesha, and is intended for the combined heat and power generation or only electricity generation. A reciprocating internal combustion engine is a gas spark-ignition four-stroke engine cooled by water, in which cylinders are arranged in the shape of a V; it is supercharged by two exhaust turbochargers and an intercooler of filling material consisting of a gaseous fuel and air prepared in the mixer.

The gas regulation network is equipped with a gas pressure regulator. The engine is equipped with an electronic ignition system and sparking plugs with platinum electrodes. The nominal mechanical engine power is 550 kW at 1500 revolutions per minute.

The electric generator Leroy Somer LSAC 49.1, whose output is 653 kVA and is located in the cogeneration unit, is powered by motors and the heat from exhaust fumes and coolant is utilized by means of a heat exchanger. The exhaust gases are discharged into a separate chimney.

The thermal module comprises heat exchangers, pumps, an exhaust silencer and an electric switchboard with a control and power part. Another part of the cogeneration unit is an oil tank allowing automatic oil refilling. The container is equipped with doors on both sides of the container and at the front in order to enable the service officer to access the propulsion unit. The container is also equipped with a removable roof allowing the system to be dismantled when necessary.

Figure 9 Cogeneration unit motorgas TBG 520

Connecting the end-user to electrical power is done through switchboard, to which a kiosk-type transformer substation is connected. The kiosk transformer substation 2 by 800 kVA is mounted on a paved surface. The transformer substation is mounted 4 metres from the existing mast substation (the protection zone of the kiosk transformer substation). The LV and HV cable link, which is laid in the ground at the depth of 1 metre, is connected to the transformer substation. The transformer substation is a building with bottom dimensions of 5.4 by 3 meters. In Fig. 10 there is a block diagram of landfill gas collection and treatment at the landfill in Úholičky.

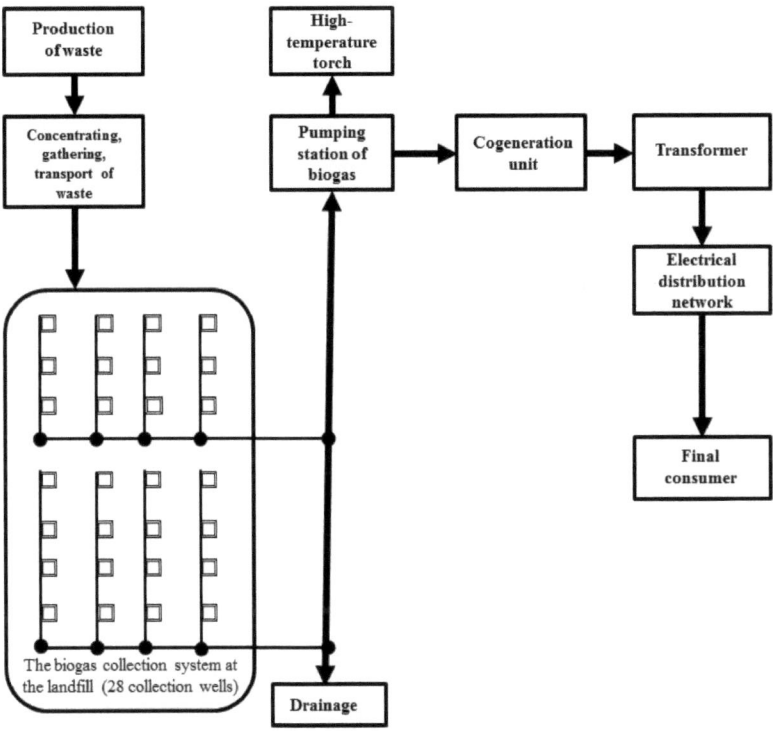

Figure 10 Diagram of landfill gas collection and treatment

The inspection shaft is used for controllable discharging of the inflowing dump water into the existing leachate reservoir and recirculation of the leachate to the landfill body. Due to the operational reasons, the inspection shaft is designed as an open object with safety elements.

4.3 DUMPING SITE OF THE LANDFILL SITE IN ÚHOLIČKY

The dumping site consisting of loess soils is located at the entrance to the complex. The dumping site D1 is situated further from the entrance to the landfill site (further south) and the dumping site D2 is located opposite the building (further north).

The dumping site D1 is 15 to 16 m higher than the surrounding terrain and its surface is not covered with vegetation. The dumping site D2 follows the dumping site D1 and the soil was deposited northwards; its height is 6 to 8 m above the ground level. The surface of the site D2 is vegetated by naturally seeded vegetation. The total volume of both sites was set by geodetic measurements at 129 000 m^3. Of this volume 108,000 to 111,000 m^3 of soil can be considered to be utilised for mineral insulation of the landfill and approximately 20,000 m^3 of soil for landscape modifications.

4.4 ADDITIONAL FACILITIES IN THE AREA OF THE LANDFILL SITE IN ÚHOLIČKY

There are a few additional operation facilities in the area of the landfill site, including an A.S.A. production line of solid alternative fuel (SAF). It is used for crushing specially separated waste, which can be used in the case-hardening furnaces in the landfill site. The maximum production capacity of the line is 10 000 tonnes of alternative fuel per year; the line does not pollute the air. The waste is processed on the line by means of two-stage crushing and the final product is handed over to the interim storage.

Moreover, operation of a new sorting line for secondary raw materials supplied by ASA, spol. s.r.o. was officially launched in April, 2012. The line capacity will enable the company to process at least 15,000 tons of paper and plastic materials each year. Furthermore, there are ongoing collection and purchase of hazardous waste, whereas the waste is stored in the hazardous waste storage site.

5. METHODS FOR MAKING THE LANDFILL GAS GENERATION EFFECTIVE

For comparison of a number of existing gas wells and the model ones and for purpose of making the operation of cogeneration unit effective by means of measuring, the following methods were selected:

> ➢ Computational simulation,
> ➢ Experimental method

Before the measurements were carried out, a mathematical model was created and it was, then, verified by simulation. Measurements of landfill gas composition and procedures for gradual decommission of gas wells were done in June, 2014. Gradually, wells 1,7,29,33,31,4,5,25,22,19 were decommissioned. The measurements were carried out by means of the Flame Ionization Detector method (FID).

6. THE METHOD OF COMPUTATIONAL SIMULATION FOR MAKING LANDFILL GAS GENERATION EFFECTIVE

The mathematical model of the landfill was created by means of two software tools. Autodesk Inventor Professional, in which the particular layers were drawn and the slope of the terrain based on the technical drawings, was used to sketch geometry of the landfill site. After the geometry was created, the entire model was imported in the computer-aided engineering software - Autodesk Simulation CFD. The geometry of model and the individual gas wells (GW) was verified in the first step. The basis of every analysis is creation of mesh. The mesh of the landfill site and the close surroundings of the gas wells is shown in Fig. 11.

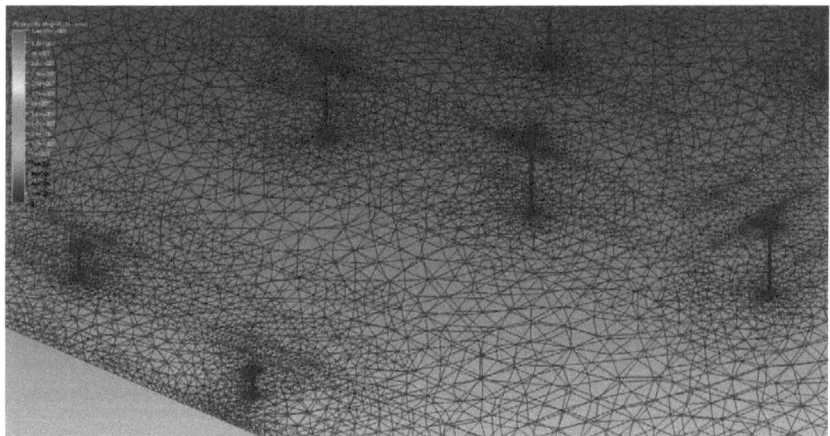

Figure 11 Mesh of the model with gas wells

When the mesh was created and checked, boundary conditions for individual parts of the model were set. Since extraction of landfill gas from a relatively large volume is analysed in this case, the analysis is of fluid flow type; particles of landfill gas are mixed.

As an illustration, simulation of a single probe, whose velocity distribution and visualization are illustrated in Fig. 12 a, was performed in a given volume. Behaviour of particular gas particles during suction was observed.

Out of the total volume, a volume consisting of two gas wells was selected, whereas behaviour during suction was observed like in the case of one gas well (Fig. 12 a). For better understanding, there are streamlines for one gas well in Fig. 13 (Fig. 13a) and for both gas wells (Fig. 13b).

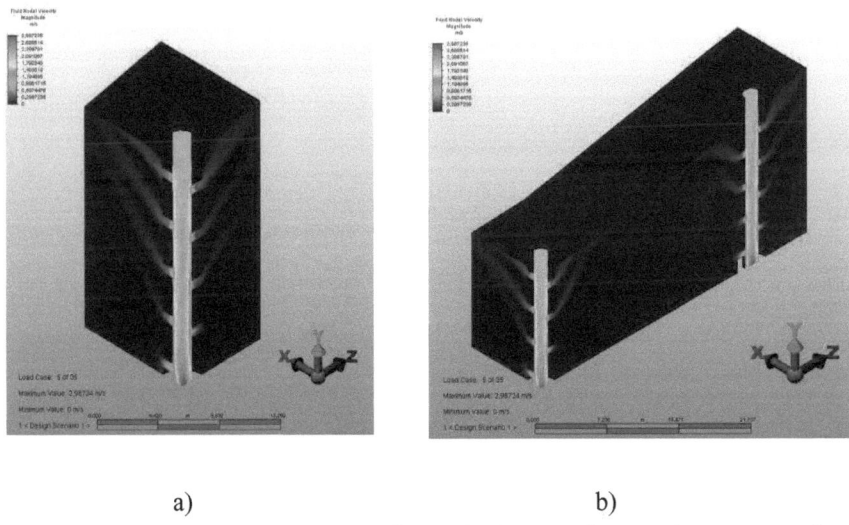

a)   b)

Figure 12 Display of extraction of landfill gas in the case of one and two gw out of a certain volume

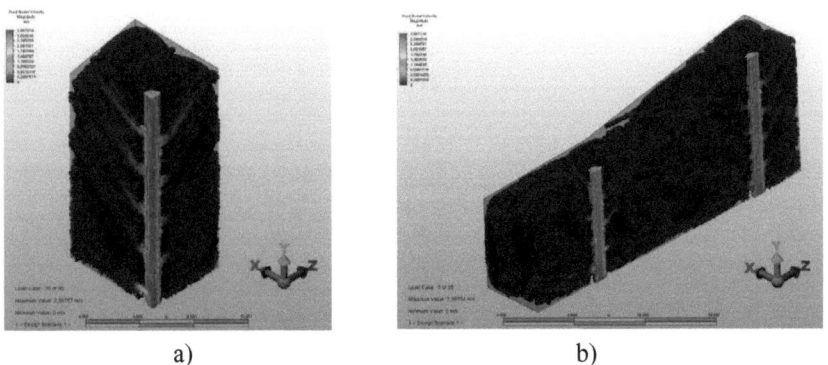

a)   b)

Figure 13 A cross-section of two gas wells together with streamlines

In Fig. 14 and 15 there are the results of flow simulation in the form of streamlines. The purpose of the simulation and the entire flow analysis at the boundary conditions was obtaining the ability of the landfill site to suck off landfill gas at lower number of gas wells in operation. Each gas well has its critical point of flow velocity set at 2 m.s^{-1}, whereas it will not be able to extract the landfill gas at lower velocity.

The first case illustrates streamlines when 28 gas wells are in operation (Fig. 14). In the second case (Fig. 15), the boundary conditions were set so that 10 gas wells were gradually eliminated during operation. After decommission of the tenth gas well, the landfill site started to show values which indicated that the service station would no longer be able to extract landfill gas of the required quality.

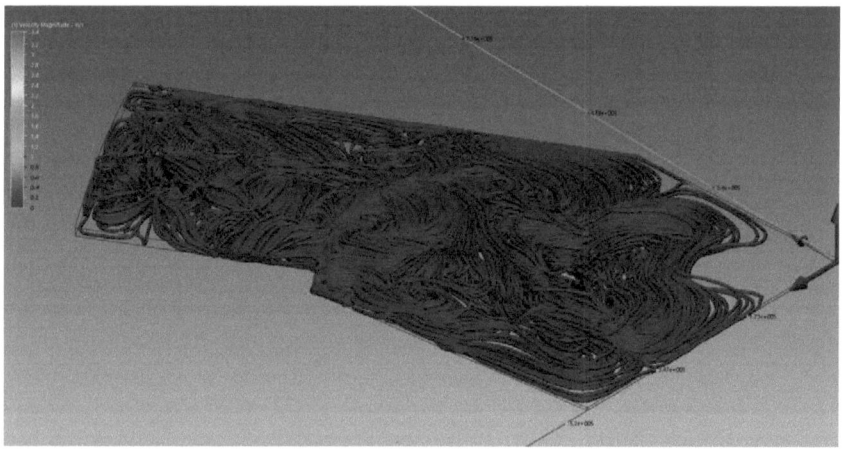

Figure 14 The recultivated part of the landfill when operating at full capacity of gas wells

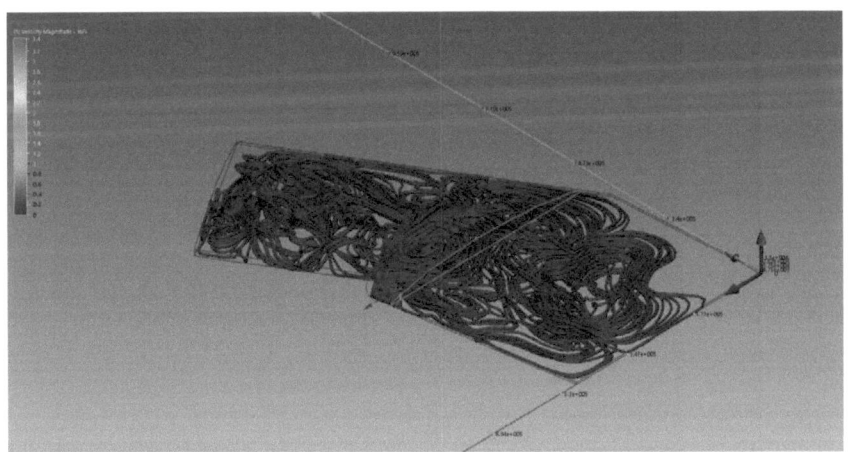

Figure 15 The recultivated part of the landfill when the gas wells are decommissioned

The results can be used for drawing up preliminary estimates and overall stability for various landfill sites if there is a recultivated part of the site and the known number of gas wells.

7. THE EXPERIMENTAL METHOD FOR MAKING LANDFILL GAS GENERATION EFFECTIVE

There are 28 active collection wells in the recultivated area of the landfill site and the collection pipeline is routed along the surface of the landfill body. The active degassing well composes of well centerings, in which there is a perforated pipe of HDPE DN 160 submerged by gravel. The cross section of a gas well located at the landfill in Úholičky is shown in Fig. 16. The degassing wells are mounted 500 mm above the bottom of the landfill.

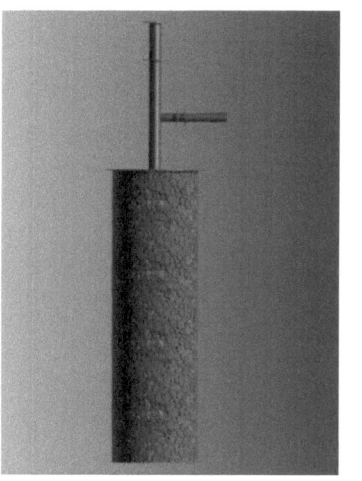 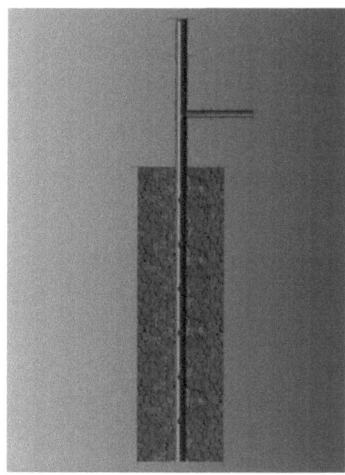

Figure 16 A gas well located at the landfill site in Úholičky and its cross section

The terrain of the area intended for the transported waste was adjusted to the needs and the pipeline was covered with soil. The common pipeline of PEHD 160 mm in dimension is held on the recultivated layer. The stand-alone connections are routed from the collection wells through the recultivated layer. They are laid on a surface of the proper inclination and in the end, they join the common gas pipeline in the site body and the main gas assembly situated at the heel of the landfill.

The leaks located outside the reclamation zone are held in a sand backfill. The resulting condensates are drawn off into adjacent wells. A regulation shaft of biogas comprises of butterfly valves which allow regulating the gas supply to the service station from the wells outside the recultivated area. The landfill gas is discharged from the main gas pipeline to the service station.

7.1. THE OBJECTIVES OF LANDFILL GAS MEASUREMENTS

The objective of landfill gas measurements carried out at the recultivated part of the landfill was to find out:

➢ The percentage composition of the landfill gas components in individual wells

➢ Streamlining of CGU operation, thus identify the critical number of wells

The measurements of landfill gas composition and gradual decommissioning of the gas wells were carried out in June, 2014. Gas wells (GW) no. 1, 7, 29, 33, 31, 4, 5, 25, 22 and 19 were gradually decommissioned. The measurements were carried out by means of the Flame-Ionization-Detector method (FID). Methane (CH_4), carbon dioxide (CO_2) and the temperature of landfill gas were measured. The meteorological parameters during the measurements are displayed in Table 11.

Table 11 The meteorological parameters during the measurements at the landfill in Úholičky

Date	16.6.	17.6.	18.6	19.6.	20.6.	23.6	24.6	25.6.	26.6.	27.6.
Temperature (°C)	22	23	25	25	20	19	21	15	18	22
Air pressure (hPa)	1013	1019	1020	1015	1014	1021	1017	1010	1016	1017
Wind speed (km.h^{-1})	3	10	5	11	16	13	7	9	13	8
Notes							rain			

The concentrations of methane and carbon dioxide in particular gas wells were obtained by measurements. The yield was 207 $m^3.h^{-1}$ of gas generated on the first day of measurements. There is the ratio of methane to carbon dioxide for each well on the first day of measurements in Fig. 17.

41

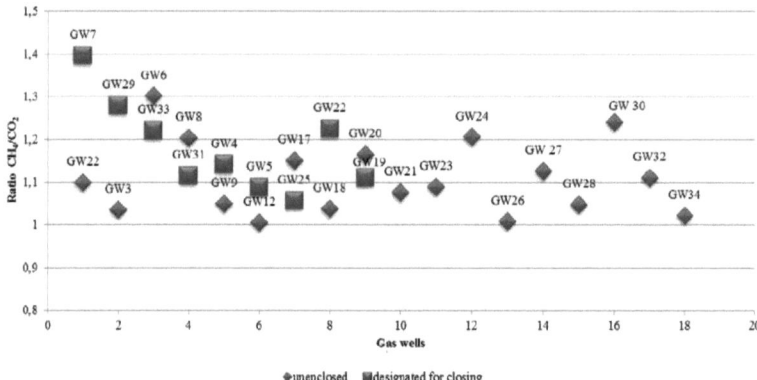

Figure 17 The ratio of methane to carbon dioxide in individual wells on the first day of measurements

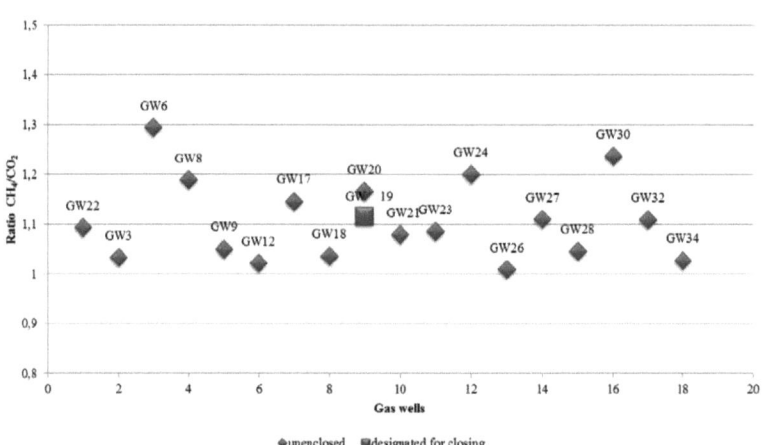

Figure 18 The ratio of methane to carbon dioxide in individual wells on the last day of measurements

The higher the ratio of CH_4 to CO_2, the better it is for the gas well. The highest ratio of methane to carbon dioxide, 1.395, was measured in gas well no. GW 7 on the first day.

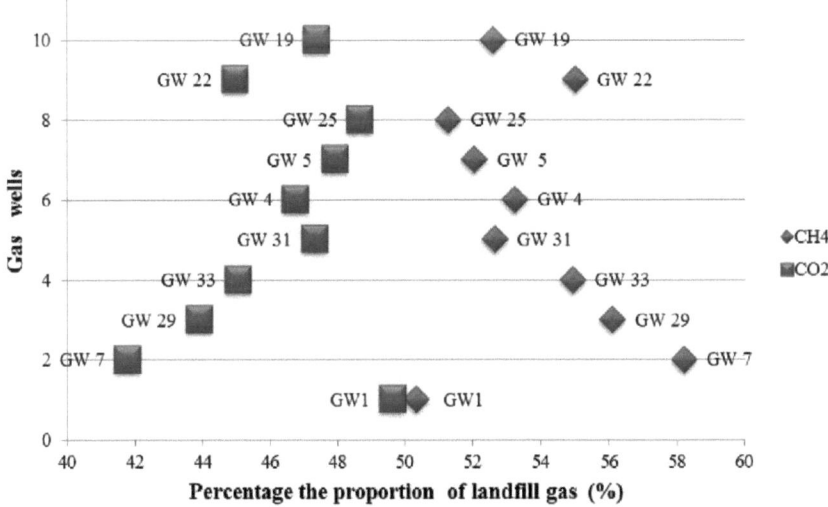

Figure 19 Graphical presentation of the components of landfill gas in the wells designated for decommissioning

The highest yield, accounting for 236 $m^3.h^{-1}$, was measured on the last day. The ratio of methane to carbon dioxide for each well after decommissioning of gas wells no. GW 1, GW 7, GW 29, GW 33, GW 31, GW 4, GW 5 and GW 25 on the last day of measurements is in Fig. 18. The highest ratio of methane to dioxide, 1.29, was recorded in gas well no GW 6 on the last day.

The results of measurements on the first day are shown in the graph in Fig. 19. The graph illustrates the measured concentrations of components of individual wells designated for decommissioning. The highest value of percentage concentration of methane measured in gas well no. GW 7 was 58.21 % and the lowest value of methane was measured in gas well no. GW 1 and it was 50.35 %.

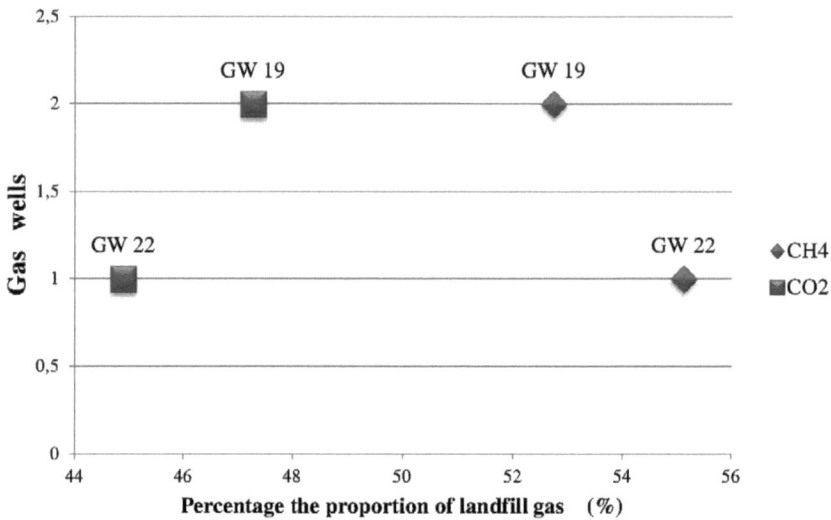

Figure 20 Graphical display of the landfill gas components measured in wells after decommissioning of wells no. GW 1, GW 7, GW 29, GW 33, GW 31, GW 4, GW 5 and GW 25

The results of measurements after decommissioning of wells no GW 1, GW 7, GW 29, GW 33, GW 31, GW 4, GW 5, and GW 25 on day 9 are displayed in Fig. 20. The highest value of methane, 55.11 %, was measured in gas well no. GW 22 and the lowest value of methane, 52.73 %, was measured in gas well no. GW 19. The lowest value of carbon dioxide measured in gas well no. GW 22 was 44.89 % and the highest value, 47.27 %, was measured in gas well no. GW 19. The highest net calorific value of 19.79 MJ.Nm^{-3} was measured in gas well no. GW 22 and the lowest value of 18.94 MJ.Nm^{-3} was in well no GW 25.

The graph in Fig. 20 shows a declining trend when gas wells no. GW 1, GW 7, GW 29, GW 33, GW 31, GW 4, GW 5 and GW 25 were gradually decommissioned. The values of methane and landfill gas flows were read from the service station on a daily basis.

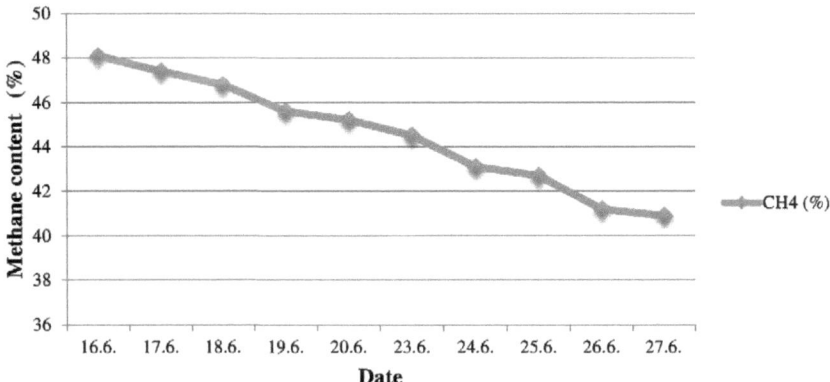

Figure 21 The content of methane at the service station

After the last gas well, GW 19, had been decommissioned, the operation of the cogeneration unit was intermitted on the same day and the value of methane recorded at the service station was 40.9%.

7.2. THE GASEOUS EMISSIONS FROM THE COGENERATION UNIT

The measurement of gaseous emissions in the cogeneration unit was carried out during ten days when the gas wells were gradually concluded. The combustion equipment burnt biogas derived from the landfill of the municipal solid waste in the complex landfill. The aim of the measurement was to measure the concentration of pollutants and monitor the impact of the number of wells on emission formation during the combustion of landfill gas in the cogeneration unit.

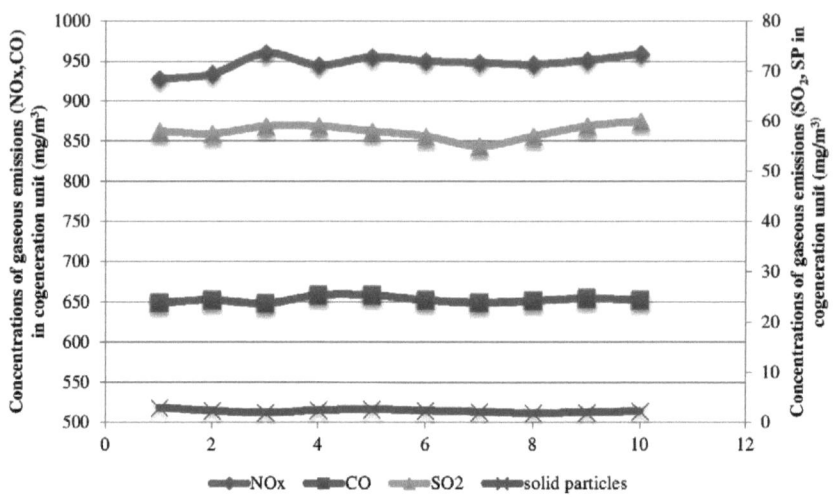

Figure 22 Average concentrations observed in emission components under normal
conditions in the dry state converted to reference content 5% O_2

The emission limit for nitrogen is 1000 mg.m^{-3}, the carbon monoxide is 1300
mg.m^{-3}, the solid particles is 130 mg.m^{-3}. For sulfur dioxide there does not exist any
fixed emission limit. The Graph in Figure 22 shows the average concentrations
observed in emission components. As can be seen in Figure 22, the number of gas
wells has no significant impact on emissions.

7.3. ELECTRICITY GENERATED IN A CHP PLANT AT THE LANDFILL IN ÚHOLIČKY

Electricity generated in the CHP plant at the landfill in Úholičky is used for
covering the own consumption of the plant and selling the electricity surplus to the
electric power transmission network. The electricity supply to the distribution
network during the measurements and the own electricity consumption of the landfill
site can be found in Fig. 23.

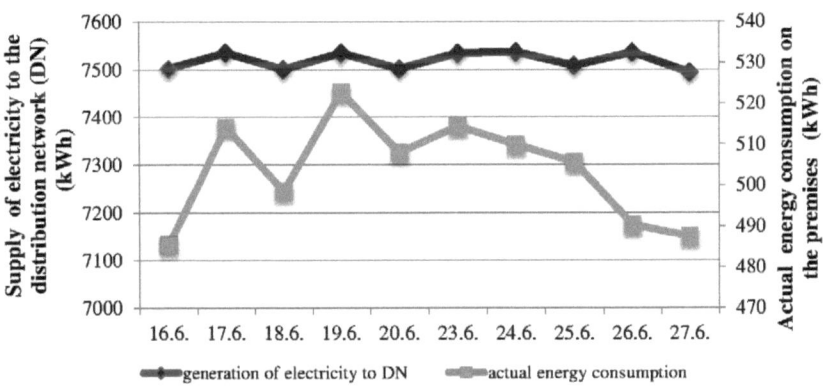

Figure 23 Electricity generated in the chp plant during the measurements

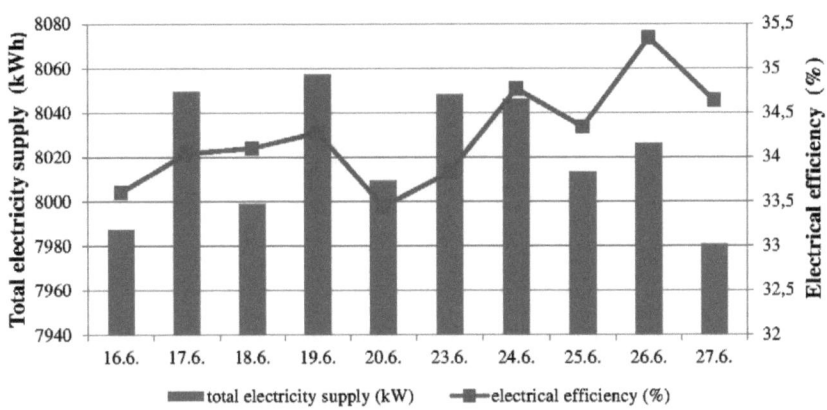

Figure 24 The total electricity supply and the degree of efficiency of electricity generation during the measurements

The graph (Fig. 24) shows the overall electricity supply, which ranged from 7981 to 8057.37 kW during the measurement. In addition, it shows efficiency of electricity generation without thermal energy recovery, which ranged from of 33.6 % to 35.34 %.

8. A PROPOSAL FOR UTILISATION OF THE RECULTIVATED AREA OF THE LANDFILL

The A.S.A line, which is designed for production of solid alternative fuel (SAF) used in the case-hardening furnaces by crushing specially separated waste, is located inside the landfill site's complex. The SAF production hall is located near the recultivated part of the landfill. It is the optimal distance for commissioning of a photovoltaic power plant (PVP), which should be located in the south of the recultivated area of the landfill. The total estimated installed power is 30 kW. All the electricity would be consumed by the SAF hall itself (Fig. 25).

The description of the technical solution:

The CHP plant currently runs on landfill gas, which is extracted from the landfill and combusted in a gas engine of the L36 Waukesha GLD type. As it is assumed that the designated landfill will be exploited with regard to landfill gas, a back-up source of electric power, in particular a photovoltaic plant of 30 kW output, has been designed to replace the CHP plant in the recultivated part of the landfill in the future.

The system consists of 120 IBC Solar Polysol 250 polycrystalline modules of 250W output placed in the open area of the recultivated part of the landfill (the south side). The most common panel size is 1590 by 980 mm. The panels will be kept in a structure made of galvanized steel (Fig. 27), which is anchored to the ground by means of screws. The individual panel structures form so-called fields making up of 6 to 12 panels per one structure; two panels above each other organized in height. The scheme of the PVP at the landfill in Úholičky is shown in Fig. 25. The 3D view of the PV panels placement located in the south of the recultivated part of the landfill is shown in Fig. 28.

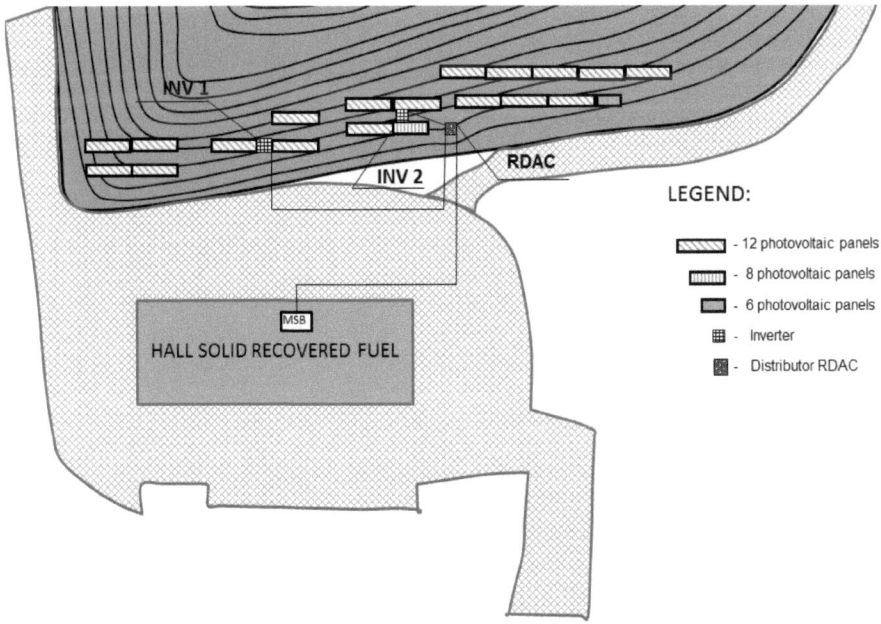

Figure 25 The scheme of the photovoltaic facility - situation - landfill in Úholičky

The DC power is supplied to two three-phase inverters, for example SMA STP 15000. It changes power to AC, which is supplied to the main switchboard after having been grounded. The inverters are of exterior design, which are usually attached to the structure below the photovoltaic panels (not exposed to the direct sunlight). The wiring is routed below the panels and far from the inverters to the RDC box, in which there are circuit-breakers, surge arresters and the power outlet to the complex.

Other elements of the connection:

➤ The outlet of the PVP is protected by an over-current safety fuse, while the main contractor, which is placed in the RAC box and connected to the existing main circuit-breaker device, serves as the main uncoupling point.

➤ A secondary electricity meter for measuring the total electricity generation at the peak of the device owned by the customer will be connected to the RAC box. In addition to this, a grid fuse (undervoltage, overvoltage, underfrequency, overfrequency, asymmetry), which has impact on the main uncoupling point in case of situation under non-standard conditions.

Figure 26 A structure of the photovoltaic panels placed in an open area (front view)

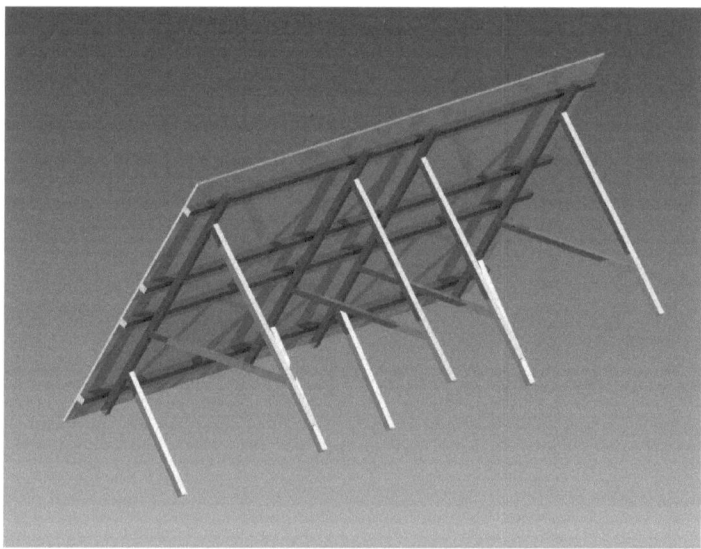

Figure 27 A structure of the photovoltaic panels placed in an open area (top view)

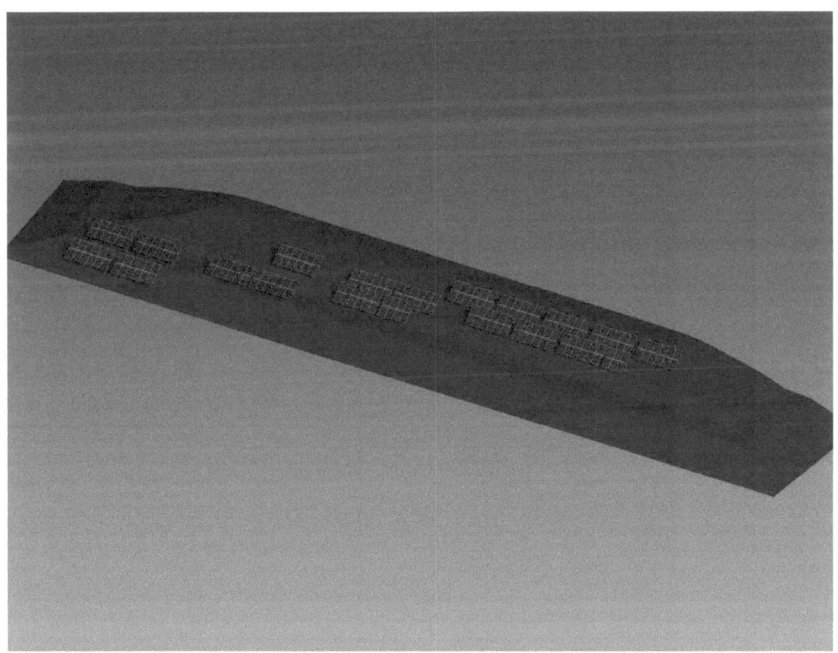

Figure 28 A view of the placement of the pv panels placed in the south of the recultivated part of the landfill

8.1 AN ECONOMIC ASSESSMENT OF THE PROPOSAL FOR UTILISATION OF THE RECULTIVATED PART OF THE LANDFILL IN ÚHOLIČKY

In order to asses the economic side of the proposal for utilisation of the recultivated part of the landfill in Úholičky, it is necessary to determine the prices of the particular technical facilities.

The description of the investment costs:

➢ **The cost of the panels.** The cost of the panels constitutes the largest part of the total investment costs. However, it is important to take into account the fact that not all the comparable panels move within the same price range. The brand, technical specifications and certificates are those that matter in this area.

➢ **The cost of the inverter.** This concerns the cost of a three-phase inverter.

- ➢ **The cost of the open area's structure** The cost of the structure for attaching the panels to the open area includes the structure itself, special ground screws made of galvanized steel, steel or aluminium, and installation of the ground screws by machines.
- ➢ **The installation cost** The installation cost makes up of the cost of the panels, installation of inverters, laying the DC and AC wiring, grounding.
- ➢ **The inspection cost** The inspection cost includes the cost of inspection report drawn up by an inspection technician and the cost of devices testing.
- ➢ **The cost of electrical installation material** The cost of electrical installation material contains the cost of one-way wiring (a cable for solar panels, 6 mm), circuit-breakers, over-current safety fuse, fuse switch-disconnector, switchboard and small connecting electrical material.

The economic assessment of the proposal for utilisation of the recultivated part of the landfill in Úholičky includes an assessment of investment and operating costs (Tab. 12 and Tab. 13).

Table 12 The budget of the particular PVP items - the landfill in Úholičky, 30kW

Item	Unit price in € without VAT	Quantity piece	Price per item in € without VAT
Panels IBC Polysol 250W	130	120	15600
Three-phase inverter SMA STP 15000	2550	2	5100
The construction of an open area	7500	set	7500
Installation	1	set	3000
Revision, testing system	1	set	1000
Electric installation material			1200
TOTAL			**33400**

Table 13 An estimate of revenues and return on investment for the output of 30 kW

Investment costs with VAT in €	40 080,00
Assumed annual production power energy in kWh of installed kWp	1066
Gross annual yield in € for the entire PV plant with a consumption of 100% of total production	3264,00
Gross return on in years	12,28
Installed capacity in W	30 000,00
Assumed annual production of PVP in kWh	32 000,00

The cost of purchasing the panels, the cost of inverters and structures, the cost of installation, inspection and electrical installation material were added to the sum of the total investment costs. The selected polycrystalline panels IBC Polysol 250W of the German technology, 12 pieces in total, represent an investment worth 15,600 Euros. The extension of the costs was directly provided by the supplier. Furthermore, it was also necessary to include two sets of three-phase inverter SMA STP 15000 worth 5,100 Euros into the investment costs.

Another item of the investment costs is the structure attached to the open area together with grub screws mounted into the soil, and it was estimated at 7,500 Euros. The investment costs of installation (panels and inverters), the costs of laying the DC and AC wiring, grounding represent 3,000 Euros. In addition, it is necessary to include the investment costs of inspection, testing and reboot of the system. The last item on the investment costs list is the electrical installation material - circuit-breakers, surge arresters, fuse switch-disconnector, wiring, switchboard and small connecting wiring material - estimated at 1,200 Euros. The total investment costs are, then, 40,080 Euros.

This chapter addresses the estimate of the PV plant's profitability in the start-up. The estimates are sufficiently accurate for small and medium-sized PV systems. SolarGIS is a high-resolution climate database operated by Solar GeoMODEL Solar. The primary data layers include an air temperature, terrain and the solar irradiation. The air temperature at a depth of 2 meters is derived from the CFSR and GFS data. The increased spatial resolution level of data corresponds to the variability of the terrain. The solar irradiation is derived from the satellite and atmospheric data.

8.2.1. THE TERRAIN HORIZON AND THE DAY LENGTH

The path of the Sun across the sky over a year for the area of landfill in Úholičky is presented in Figure 29. The terrain horizon (in gray) and the horizon of the panels (in blue) can cause shading and decrease the amount of the incident solar radiation.

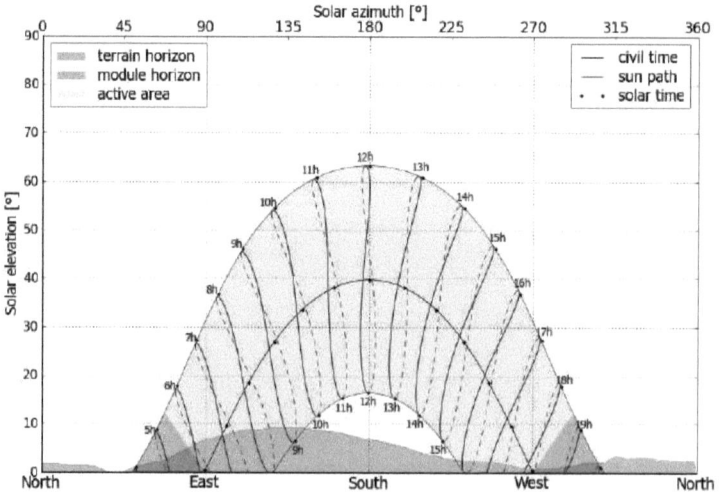

Figure 29 The path of the sun over a year

Black dots indicate the local solar time. The blue numbers in the graph indicate the Central European clock time.

8.2.2. THE GLOBAL HORIZONTAL SOLAR IRRADIATION AND AIR TEMPERATURE - THE CLIMATE REFERENCE VALUES

The monthly sum of global irradiation (Gh_m), the daily sum of global (direct) irradiation (Gh_d), the daily sum of diffuse irradiation (Dh_d) and the daily air temperature (T_{24}), therefore, the climate reference values are listed in the table. The daily sums of global and diffuse irradiation together with the air temperature over 24 hours are displayed in Fig. 30.

Table 14 The climate reference values

Month	Gh_m (kWh.m^{-2})	Gh_d (kWh.m^{-2})	Dh_d (kWh.m^{-2})	T_{24} (°C)
Jan	27,5	0,89	0,54	-1,0
Feb	43,9	1,57	0,89	0,4
Mar	81,4	2,63	1,48	3,7
Apr	125,9	4,20	2,11	8,1
May	157,8	5,09	2,57	13,1
Jun	163,5	5,45	2,79	16,4
Jul	159,8	5,16	2,64	18,6
Aug	142,3	4,59	2,29	18,8
Sep	95,1	3,17	1,65	13,9
Oct	57,8	1,86	1,07	9,0
Nov	27,8	0,93	0,60	3,5
Dec	19,5	0,63	0,41	-0,4
Year	1102,3	3,02	1,59	8,7

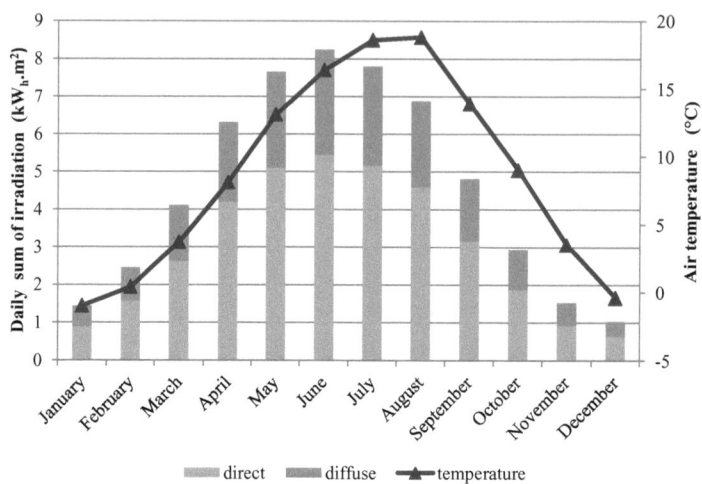

Figure 30 The daily sums of direct and diffuse irradiation together with the air temperature

8.2.3. THE GLOBAL IN-PLANE IRRADIATION

The losses of the global irradiation by terrain shading (Sh_{loss}), the daily sums of reflected irradiation (Ri_d), the daily sums of diffuse irradiation (Di_d), the daily sums of global (direct) irradiation (Gi_d) and the monthly sums of global irradiation (Gi_m) are shown in the table. The daily sums of reflected irradiation, the daily sums of diffuse irradiation and the daily sums of global irradiation are shown in the graph in Fig. 31.

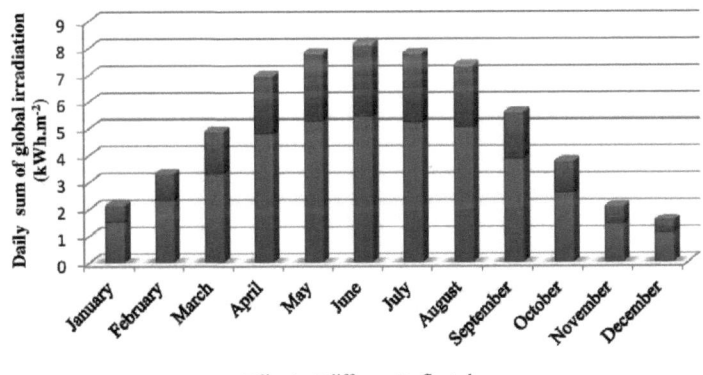

Figure 31 The daily sums of reflected irradiation, the daily sums of diffuse irradiation and
the daily sums of global irradiation

8.2.4.   *THE PV ELECTRICITY PRODUCTION IN THE START-UP*

The monthly sums of specific electricity production (Es_m), the daily sums of
specific electricity production (Es_d), the monthly sums of total electricity production
(Et_m), the percentual shares of monthly electricity production (E_{share}) and the
performance ratio (PR) are illustrated the graph in Fig. 32.

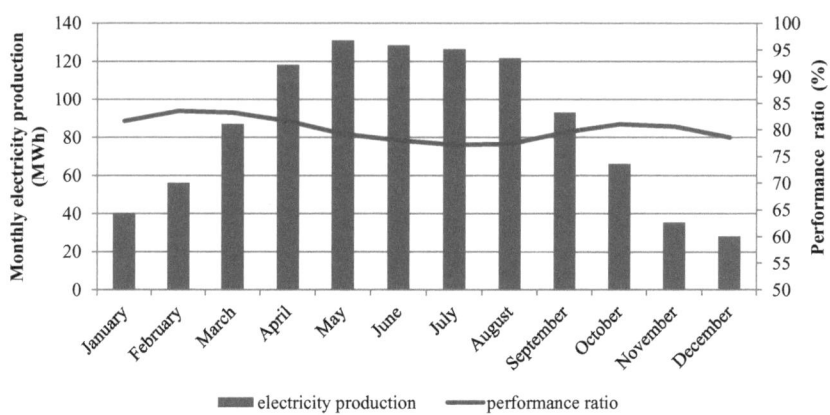

Figure 32 The estimated annual electricity production and efficiency of the system

As can be derived from the graph in Figure 32, the highest electricity production, 130.9 MWh, will be in May and the lowest electricity production will be in December accounting for 28.1 MWh. The effectiveness over the year will move within a range of 79.6% on average.

The summary of the results of the expected annual electricity generation
The annual global in-plane irradiation: 1269 kWh.m^{-2}
The average annual air temperature at 2 m: 8.7 ° C
The average annual electricity production: 32 MWh

9. THE ASSESSMENT OF THE OBTAINED RESULTS AND THE PROPOSAL FOR UTILISATION OF THE RECULTIVATED PART OF THE LANDFILL

At present, landfill gas is not only used for power generation in the world but also in Slovakia. In order to utilise landfill gas in respect to energy, it is necessary to know the course of its generation. Economic valuation of energy recovery from landfill gas generated at the landfills depends on the expected amount of flue gases. The practically usable quantities of landfill gas are much lower. [59] The amounts of methane generated from solid municipal waste and the theoretically possible heat and power generation in a CHP system were found in Slovakia on the basis of the theoretical calculations. The majority of electric and thermal energy would be generated in the Nitra region, and the lowest amount would be generated in the Bratislava region.

It is very important for the operator of the cogeneration unit to prevent the gas wells from malfunctioning. If a similar situation occurs, information on the critical amount of wells is very valuable in this scenario. This is also important in the economic sense of terms because the operator loses revenues for each day of malfunctioning of the system. When the content of methane contained in landfill gas is at 40%, operation of the cogeneration unit is ceased. Therefore, it is important for the content of methane not to decrease below 40 % so that the efficient operation is maintained.

Before the measurements were carried out, a mathematical model was created and, simultaneously, it was verified by simulations. The one- [60] or two-dimensional [61] models have been developed so far. The model generated for the recultivated part was three-dimensional (3D). Since the recultivated part of the landfill was examined, the selected model included impermeable boundary conditions. Until now it was assumed in the computational models that the pressure inside the gas wells was atmospheric [61], however, the pressure in the gas wells is not atmospheric in practice, which the model took into account. The most important fact was that the critical value of methane content was reached after gradual decommissioning of the gas wells, in particular ten wells. This was how the critical number of gas wells by means of numerical simulation was obtained.

The actual measurements carried out at the landfill confirmed that the methane content at the service station reached the value of 40.9 % after ten gas wells had been decommissioned. On the same day, decommissioning of the tenth well caused intermittence of the cogeneration unit's operation. In actual fact the cogeneration unit ceased to operate when the tenth well was decommissioned. Thus, the real-time measurements confirmed the simulation results. In the end, we can conclude that the critical number for a landfill containing 28 active gas wells is 10 wells.

Based on the results of the measurements carried out, it can be concluded that the landfill site was in the middle of phase IV of landfill gas generation at the time of measurements. [14, 16]. In this phase, the gas generation is of relatively constant nature. The methanogenic-steady process takes place exclusively under the anaerobic conditions and with methanogenic bacteria present in the process. The main gaseous product of this phase is methane and carbon dioxide, which was confirmed by the results of the measurements. The measured values of methane ranged from 50.12% to 56.58%. And the measured values of carbon dioxide ranged from 43.40% to 49.89%. The measurements confirmed that the ambient temperature does not have any significant impact on methane generation [7]. The ideal ratio of methane to carbon dioxide was 1.12. The higher the ratio of CH_4 to CO_2, the better for the concerned gas well. On the first day, the highest ratio of methane to carbon dioxide was 1.395 and was measured in well no GW 7, while the highest ratio of methane to carbon dioxide, in particular 1.29, measured on the last day was recorded in well no. GW 6. The yield measured on the first day accounted for 207 m3.h-1 and the highest yield, 236 $m^3.h^{-1}$, was reached on the last day of measurements. The average efficiency of electricity generation without thermal energy recovery during the measurements was 34.24%.

The monograph deals with utilisation of the recultivated part of the landfill after exhausting landfill gas. A PV plant of the total estimated installed capacity of 30 kW was designed in this area. The system consists of 120 IBC Solar Polysol polycrystalline modules of 250W output placed in the open area of the recultivated part of the landfill. The estimated annual electricity generation at the landfill in Úholičky involves an estimate of the profitability of the PV plant in the start-up. This estimate takes into account 365 days a year. The highest electricity generation would be in May and it would represent 130.9 MWh. On the other hand, the lowest electricity production would be in December, 28.1 MWh. The efficiency would move within a range of 79.6 % on average throughout the year.

10. Contribution of the obtained results and proposed solutions to the addressed issue

The contribution of the monograph to the addressed issue lies mainly in the possibility of using the results and proposals referred to therein. It summarizes current knowledge of generation and utilisation of landfill gas as well as the calculations of the potential heat and power generation in a CHP plant in the Slovak Republic.

The theoretical contribution to the theme is represented by thorough data processing on the landfill sites and gas in the Slovak Republic; the data was taken from credible domestic and foreign sources. Consequently, the lessons learnt in the context of theory in solving the issues mentioned in this monograph may be applied in the educational context in the future.

Another contribution to the educational activity is acquiring the skills necessary for creating 3D models and computational mesh in Autodesk Simulation CFD. At the same time, the evaluation of experimental measurements is another important contribution to the educational process.

Contribution to science and research is the mathematical model, which can be used to illustrate the landfill gas flow at different inlet parameters. This means that there is a possibility of changing the parameters and conditions of the degassing system of the landfill site in the software. Dissemination of the obtained results and knowledge by means of publication, conferences and lectures can be considered to be additional contribution of the monograph.

Contribution to practice is a practical guide to more effective operation, that is to say, knowledge of the critical number of the gas wells necessary for operation of a CHP plant. In addition to this, the book has contributed to proposing another method of utilisation of the recultivated part of the landfill after depletion of the gas.

CONCLUSION

The landfill gas generation procedure is a well-known fact and, at the same time, a topical mater that is currently dealt with in various scientific research assignments. Landfill gas is a sort of biogas produced during anaerobic digestion of organic materials that are disposed of at a landfill, mechanically compacted and gradually applied in layers a few meters thick. Its generation mostly depends on the sort of material disposed of at the landfill and the decomposition stage of the organic matter disposed of at the landfill. The composition of landfill gas varies according to the speed of exhausting and the age of the landfill.

From the standpoint of security and environment, it is necessary to degas the landfills and dispose of the landfill gas or secure its energy recovery.

The aim of the monograph was to study the process of landfill gas generation and its use in a CHP plant at a landfill. Combustion of landfill gas in a CHP plant is not an unknown technological procedure and this method of waste treatment has already been used at eight municipal waste landfills in the Slovak Republic.

The method of theoretical calculation of landfill gas production applied to the landfill sites in Slovakia was presented in the monograph. The elementary input data on the number and composition of waste for calculations were extracted from the register of the landfills managed by the State Geological Institute of Dionýz Štúr. In spite of the results of calculations, it is necessary to carry out additional measurements for the determination of production and composition of the particular landfill gas at the designated landfill site when an operator makes a decision to utilise the gas in terms of energy.

This monograph also contains the measurements of landfill gas production at the landfill site in Úholičky for purpose of comparing the number of the existing gas wells with the model ones and consequent streamlining of operation of a CHP plant. A computational simulation for determination of the optimal number of the gas wells preceded the measurements. The measured values show that the site in Úholičky was in the middle of phase IV of landfill gas generation at the time of measurements; the main gaseous products of this phase are methane and carbon dioxide. In this phase, the amounts of generated landfill gas are high enough and relatively constant. The results of the measurements confirmed that the simulation results, which focused on finding the optimal number of gas wells in order to ensure optimal operation of the CHP plant at the landfill in Úholičky, were accurate.

Proposal for a photovoltaic power plant on of landfill body can be applicable not only to recultivated part landfill but on the old landfill which has not be used for energy production. The monograph dealt with the proposal of utilisation of the recultivated part of the landfill site in Úholičky. It goes without saying that when the landfill gas is actively drawn off, there arises a scenario in which the body (the recultivated part) will not be capable of landfill gas generation. In the light of the abovementioned, a question of utilisation of the recultivated part of the landfill arises. On the grounds of the fact, a back-up energy source, a photovoltaic plant of designed power of 30 kW, was proposed to replace the original energy source in the recultivated area of the landfill. This proposal can also be used for the old landfill sites which no longer serve their purpose - energy generation.

The study could further pay attention to finding an optimum placement of the gas wells in the selected area of the landfill site with respect to the total yield of the wells.

REFERENCES

[1]. Statistical Office: Waste in the Slovak Republic [online 2014-12-11] , Available on the Internet <http://slovak.statistics.sk/wps/portal> .

[2]. Decree no. 310/2013 Collection of Laws of the Ministry of Environment of the Slovak Republic from September 18, 2013 implementing certain provisions of the Waste Act. Ministry of the Environment Report on the state of the environment [online 2014-12-11], Available on the Internet: <http://www.minzp.sk/dokumenty/sprava-stave-zp/>.

[3]. KLENOVČANOVÁ, A., IMRIŠ, I.: Resources and energy conversion, 1st ed. - Prešov : ManaCon, - 2006. – p.492 - ISBN 80-89040-29-2.

[4]. TAKÁČOVÁ, Z., MIŠKUFOVÁ, A.: Basic information on the waste, Košice : Equilibria - 2011. – p.223 - ISBN 978-80-89284-78-8.

[5]. HORBAJ, P., SCHVARZBACHEROVÁ, E.: The use of biogas in cogeneration, Košice 2011, p.184 ISBN: 978-80-553-0577-6.

[6]. STRAKA, F. et al .: biogas. 2nd ed. GAS s.r.o., Praha 2006. P.706 ISBN 80-7328-090-6.

[7]. FRANZIUS, L V.: Gefährdung durch Deponiegas.Müll Handbuch Kennzahl 4589 Erich Schmidt Verlag, Berlin Germany, (1981) .

[8]. WANG, X. et al: Greenhouse gas emissions from landfill leachate treatment plants: A comparison of young and aged landfill. In: Waste Management. Roč. 34, (2014), p. 1156–1164. ISSN: 0956-053X.

[9]. FERRAZ, F.M. et al: Co-treatment of landfill leachate and domestic wastewater using a biofilter. In: Journal of Environmental Management. Roč. 141, (2014), p. 9-15. ISSN: 0301-4797.

[10]. HEERTEN, G., F.M. et al: Cover systems for landfills and brownfields. In: Land Contamination and Reclamation. Vol. 16, No. 4, (2008), p. 343-356. ISSN:0967-0513.

[11]. AMINI, A.R . et al.: Comparison of first-order-decay modeled and actual field measured municipal solid waste landfill methane data. In: Waste Management. Vol. 33, (2013), p. 2720– 2728. ISSN: 0956-053X.

[12]. BOCKREIS, A., et al.: Gaseous emissions of mechanically-biologically pre-treated waste from longterm experiments. In: T. Christensen, R. Cossu, R. Stegmann (Eds.), Proceedings of the SARDINIA 2003 – Ninth International Waste Management and Landfill Symposium, Cagliari, Italy.

[13]. RETTENBERGER, G., MEZGER, H.: Langzeitphasen des Deponiegeschehens bei Altablagerungen, Industrial waste management, waste reduction and treatment; Envirotech, (1992). p. 487–494.

[14]. KRÜMPELBECK, I.: Study on the longterm behaviour of municipal waste landfills. Wuppertal, Bergische University GH Wuppertal, FB Construction Engineering, PhD thesis, (2000).

[15]. WAGNER, J.-F. et al .: Modern landfill technology–landfill behavior of mechanicalbiological pretreated waste. In: Proceedings Sardinia 2007, Eleventh International Waste Management and Landfill Symposium. Sardinia: CISA, 2007.p.50-60.

[16]. RETTENBERGER, G. : Landfill decommissioning A sensible approach to sustainable waste management. In: Stegmann, Rettenberger (hrsg.): 12, Bonn: Economica Verlag, 1998. ISBN: 3-87081-028-9.

[17]. YOUNG, P. : An Assesment of the Odor and Toxicity of the Trace Components of Landfill gas. Proc. GRCDA 8th Int. Landfill Gas Symp., 9.-11.4. San Antonio TX, USA. (1985). p.93-114.

[18]. Register of the landfills. [online 2014-12-31]. Available on the Internet: < http://www.geology.sk/new/sk/sub/ms/geof/skladky>

[19]. FÁBER A. et al.: Atlas of renewable energy sources in Slovakia, Bratislava Energy Centre, 2012, ISBN 978-80-969646-2-8.

[20]. PAWLOWSKA, M.: Mitigation of Landfill Gas Emissions, by CRC Press 2014 , p.100, ISBN 9780415630771

[21]. ČERMÁK, O.: Waste management, Landfill gas. STU Bratislava 2009. 134 p. ISBN 978-80-227-3101-0

[22]. MUNTONI, A. et al.: An integrated model for the prediction of landfill emissions. Proceedings Sardinia 95. Volume I. october 1995.p. 231

[23]. HORBAJ, P.: Theoretical calculation formation of methane from municipal waste, Chemical Letters 98, (2004), p 137 - 141, ISSN 1213-7103.

[24]. BOVE, R. LUNGHI, P.: Electric power generation from landfill gas using traditional and innovative technologies, Vol. 47, No. 11-12, (2006), p. 1391-1401,

[25]. JAFFRIN, A. et al.: Landfill Biogas for heating Greenhouses and providing Carbon Dioxide Supplement for Plant Growth, Vol. 86, No. 1, (2003), p. 113-123

[26]. Clarke Energy. [online 2015-01-05] Available on the Internet: <http://www.clarke-energy.com/landfill-gas/>.

[27]. HOLOUBEK: Combined heat and power, tri-generation and heat network. 1st ed - Košice: HF, 2008. p.159 ,ISBN 978-80-8073-977-5.

[28]. Environmental Protection Agency. [online 2015-01-05] Available on the Internet : http://www3.epa.gov/lmop/documents/pdfs/pdh_chapter3.pdf.

[29]. ZAMORANO M. et al.: Study of the energy potential of the biogas produced by an urban waste landfill in Southern Spain, , Vol. 11,No. 5, (2007), p. 909 – 922.

[30]. BIDART, Ch., FRÖHLINGA, M., SCHULTMANNA F.: Municipal solid waste and production of substitute natural gas and electricity as energy alternatives, Vol. 51, , No. 1-2, (2013),p. 1107-1115.

[31]. HAO, H. et al.: Trigeneration: A new way for landfill gas utilization and its feasibility in Hong Kong, Vol. 36, (2008), p. 3662-3673.

[32]. TSATSARELIS, T. et al. : Technologies of landfill gas management and utilization." Protection and Restoration of the Environment VIII, Chania (2006), p. 3-7.

[33]. A. JOHARI, S. I. AHMED, H. HASHIM et al.: Economic and environmental benefits of landfill gas from municipal solid waste in Malaysia, Vol. 16, 2012, No. 5, p. 2907-2912.

[34]. D. GEWALD, K. et al.: Waste heat recovery from a landfill gas-fired power plant, Vol. 16, (2012), p. 1779-1789.

[35]. FINDIKAKIS, A. N., a LECKIE, J. O. :Numerical simulation of gas Jow in sanitary landfills. Journal of the Environmental Engineering Division, ASCE, 105, (1979), p.927.

[36]. FINDIKAKIS, A. N. et al.: Modeling gas production in managed sanitary landfills. Waste Management Research, 6, (1988), p.115.

[37]. SAHIMI, M. : Flow phenomena in rocks: From continuum models, to fractals, percolation, cellular automata, and simulated annealing. Reviews of Modern Physics, 65, (1993), p.1393.

[38]. KNACKSTEDT, M. A., et. al. : Pore network modelling oftwo-phase Jow in porous rock: The effect ofcorrelated heterogeneity. Advances in Water Resources, 24, (2001), p.257.

[39]. WILKE, C. R. :A viscosity equation for gas mixtures. Journal of Chemical Physics, 18, (1950), p.517.

[40]. BIRD, R. B., et al.: Transport phenomena (2nd ed.). New York: Wiley, (2002).

[41]. HIRCHFELDER, J. O., CURTISS, et al.: Molecular theory of gases and liquids. New York: Wiley, (1954).

[42]. MORCHICK, L., and MASON, E. A. : Transport properties ofpolar gases. Physics of Fluids, 2, (1961). p.449.

[43]. YOUNG, A. : Mathematical modeling oflandfill gas extraction. Journal of the Environmental Engineering Division, ASCE, 115, (1989). p.1073.

[44]. BETUŠ, Z.: Energy Recovery biogas in cogeneration units. Žilina: Acta Mechanica Slovaca, 3/1999. p.121-126.

[45]. TRNAVSKÝ, J.: Inside the landfill waste stream out electrical current - municipal equipment 9/2009, Profi Press, s.r.o., Praha.

[46]. TAUŠ, P., TAUŠOVÁ, M.: Economic assessment of the return of selected types of RES. In: Options for financing energy projects in Slovakia and the EU:

Podbanské, 5th-7th June 2006: Proceedings of the national professional conferences. Košice: House of techniques, 2006. ISBN 80-232-0262-6.

[47]. VÍGLASKÝ, J. et al.: Landfill Gas Utilization in the Slovak Republic. Jönköping: medzinárodná konferencia WORLD BIO ENERGY, 2006. p. 361

[48]. PESTA, G. – RUß, W.: Die zuverlässige Reinigung von Biogas Verfahren und Lösungsansätze. In: Innovationen in der Biogastechnologie. Deggendorf Fachhochschule Deggendorf, 2004. p.99-123

[49]. CHRISTENSEN, Th.H., KJELDSEN, P., LINDHARDT, B.: Gas generating processes in landfills. In: Christensen, Th.H., Cossu, R., Stegmann, R. (Eds.), Landfilling of Waste: Biogas. E&FN Spon Verlag, London, (1996), ISBN: 0-419-19400-2.

[50]. WAGNEROVÁ,E.: The energy utilization of the municipal wastes heaps. Acta Mechanica Slovaca, 2, 1998, 3, p. 268-272.

[51]. TABASARAN, O.: Gas production from landfill. In: Bridgewater AV, Lidgren K, editors. Household waste management in Europe, economics and techniques. New York: Van Nostrand Reinhold Co.; 1981. p. 159–75.

[52]. COSSU, R., et al.: Biogas emissions measurement using static and dynamic flux chambers and infrared method environmental impact, after care and remediation of landfills. Proceedings Sardinia 97, Cagliari 1997, Volume IV.p.103-114.

[53]. ABICHOU, T. et al.: Characterization of Methane Oxidation at a Solid Waste Landfill, Journal of Environmental Engineering, 2005, ASCE, Vol. 132, No. 2, p. 220-228.

[54]. MORCET M. et al.: Methane mass balance: a review of field results from three French Landfill case studies - Ninth International Waste Management and Landfill Symposium Sardinia 2003, S. Margherita di Pula (Cagliari, Italia), October 6-10, 2003.

[55]. ZENKER, M. J., BORDEN, R. C., and BARLAZ, M. A.: Biodegradation of Cyclic and Alkyl Ethers Using Attached Growth Reactors. In.: J. of Environ. Eng., 130, 9, p. 926 - 31.

[56]. BOGNER, J., et al. : Field measurement of non methane organic coumpound emissions from landfill cover soil - Ninth International Waste Management and Landfill Symposium Sardinia 2003, S. Margherita di Pula (Cagliari, Italia), October 6-10, 2003.

[57]. KJELDSEN, P. et al. : Present and Long Term Composition of MSW Landfill Leachate – A Review, Critical Reviews. In: Environmental Science and Technology, 32, 4, p. 297-336.

[58]. BOHN, S., JAGER, J.: Microbial methane Oxidation in Landfill Top Covers – Process Study on an MBT Landfill.In.: XII International Waste Management and Landfill Symposium, 5th – 9th of October 2010, S. Margherita di Pula, Sardinia, Italy.

[59]. LU, A. H., and KUNTZ, C. O.: Gas-Jow model to determine methane production at sanitary land"lls. Environmental Science and Technology, 15, (1981), p.305.

[60]. ARIGALA, S. et al. : Gas generation, transport and extraction in MSW landfills. Journal of Environmental Engineering Division, ASCE, 121, (1995). p.33.

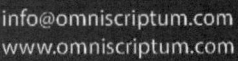

Printed by Books on Demand GmbH, Norderstedt / Germany